BIO-ENERGY FOR RURAL ENERGISATION

About the Editors

Dr. R.C. Maheshwari Ph.D. from (IIT Delhi in 1974 is professor, at Centre for Rural Development and Technology, at Indian Institute of Technology, Delhi). Prof. Maheswari has been actively persuing research, development and Extension work for the last 20 years in the area of Bio Energy Systems for Rural Areas, Extraction Technology, waste recycling and waste management, water sanitation and pulp & paper technology. He worked at the technical university, Karlsruhe, Germany in 1975-76 under the DAAD programme. He operated several research and industrial projects independently. He has published widely in National & International Journals, and Proceedings. He had been coordinator for National Technical backup unit, National programme on improved cookstoves in India from (1988-93). He is Head SCS and President Board of Student Welfare since 1996 at IIT Delhi. Presently he is a honorary secretary of Bio Energy Society of India.

Pradeep Chaturvedi, is a well known expert in the Energy Conservation and Renwable Energy sector. His expertise lies in project planning, implementation and management with emphasis on technology transfer and skill building. A number of pioneering projects and studies of national and international significance have been conducted under his guidance. He has planned, developed and implemented projects in the Asia-Pacific region. At present, Mr. Chaturvedi is also working as a Consultant to FAO's Regional Wood Energy Development Programme for Asia.

Mr. Chaturvedi is the Vice Chairman of the World Renewable Energy Network supported by UNESCO and based in U.K. Being a founder of the Bio Energy Society of India, and now its Vice President, he is inolved in projecting the bio fuels issue at national and international.

Mr. Chaturvedi is a Graduate in Engineering. He is a Fellow of the Institution of Engineers (India) and the Recipient of FAO's World Food Day Silver Medal 1991.

Mr. Chaturvedi has edited and authored 16 books related to sustainable development, and renewable energy and energy conservation: and has contributed papers at a number of national and international conferences.

BIO-ENERGY FOR RURAL ENERGISATION

(Proceedings of the National Bio-Energy Convention-95 on Bio-Energy for Rural Energisation, organised by Bio-Energy Society of India, during December 14-15, 1995 at I.I.T. New Delhi)

Edited by

R.C. MAHESHWARI
PRADEEP CHATURVEDI

CONCEPT PUBLISHING COMPANY, NEW DELHI 110059

ISBN 81-7022-670-8

First Published 1997

Published and Printed by

Ashok Kumar Mittal

Concept Publishing Company

A/15-16, Commercial Block, Mohan Garden,
NEW DELHI - 110059 (India)

Phones: 5554042, 5504042
Fax: (011) 5598898

CONTENTS

PART SEVEN
Biomethanation

PART EIGHT
Bio-Fertilization

PART NINE
Industrial and Urban Waste

PART TEN
Draught Animal

PART ELEVEN
Improved Cooking Devices

INAUGURAL ADDRESS

Col. Rao Ram Singh
Minister for Wastelands Development
Government of India

It gives me great pleasure to be present here today to discuss the theme of 'Bio-Energy for Rural Energisation' which is very close to my heart. I live in a rural area myself and am quite familiar with the needs of Bio-Energy of the rural people. I am also familiar with the great change that small scientific innovations can make in this field. Recently, I have planted eucalyptus on a farm using seedlings prepared through vegetative propagation. Not only is the biomass produced of a far greater quantity than through the normal system but the time taken to produce it is less.

The need for greater emphasis on bio-energy is because, in India there is a great diversity in living conditions varying from people still living in conditions of near stone age to those near computer age. In this context, the energy from firewood and cow-dung is as relevant as that from atomic energy. In our country, more than 76 per cent population is in rural areas of which about 40 per cent are below the poverty line. The various energy systems most relevant to the rural areas are fire wood, bio-gas, wind mills, solar devices, mini and micro hydel systems. With regard to use of bio-energy in the rural areas fire wood and bio-gas are the two most relevant sources of energy. Biomass based energy is an integral part of the overall village system which cannot be taken up in isolation.

Significance of Fuelwood

In the rural sector about 95 per cent cooking is from biomass based fuel i.e. fuelwood, dung and agricultural waste. The demand for fire wood will definitely raise to the order of about 350 million tonnes in 1996 against the present supply of about 165 million tonnes. The yawning gap between the demand and supply of fire-wood can only be met through massive afforestation programme, especially to be taken up on a vast extent of wastelands available throughout the country. The objective has to be that every village is able to meet its fire-wood, fodder and timber requirement from the land which can be easily made available for this purpose. Obviously, the only land available for this activity would be the degraded land/wasteland which is not being put to any substantial productive use. The success of this will depend on correct planting techniques, use of high yielding geno-types and fast growing varieties that are easily adaptable on these lands. Other connected aspects are:

(i) *Charcoal*: There is a need for modernisation of the manufacturing of this age old source of energy through improved research and development techniques.

(ii) *Briquetting*: This involves technology by which saw dust is converted to smokeless fuel. This area also need a thrust, so that it can be widely propagated in the Rural Sector at a low cost by using improved indigenous techniques.

(iii) *Improved Chulla Programme*: The traditional chullas are being used in the rural sector for at a very low thermal efficiency resulting in enormous wastage of energy which ultimately result in greater degradation of forest by use of more fire wood. Improved chullas are already being used which give higher thermal efficiency requiring low input of fuel, wood, dung, etc. This also needs to be popularised in a big way, so that it proliferates much more in the rural sector.

Biogas—A Solution to Rural Energisation

Bio-gas offers an important solution to the present energy crisis, especially in the rural areas. Besides being an important domestic energy for cooking, it offers an environmentally clean technology.

The enormous potential of this technology can be gauged from the fact that about 1000 million tonnes of animal dung is available annually from about 4160 million cattle population. This can produce upto 40,000 million cubic metres of gas through bio-gas plants. The slurry can produce about 400 million tonnes of organic manure which can be safely used in the agricultural fields. It has been estimated that 1,50,000 bio-gas plants can result in saving 6,00,000 tonnes of wood equivalent per year.

Great efforts have been made in developing indigenous technology for establishment of bio-gas plants in the rural sector as individual or community efforts. Lakhs of bio-gas plants have been set up all over the country, but even now there is a need to standardise the bio-gas technology, so that it can have a higher efficiency of output, at a lower cost.

Department of Wastelands Development was set up in July 1992 and was given a mandate to develop the country's non-forest wastelands on sustainable basis and to provide additional biomass to the rural people specially fuelwood and fodder. As you are aware there are about 95 million hectares of non-forest wastelands in the country. These lands are degraded due to water or wind erosion, have poor vegetative cover and are producing below their optimum capacity. My Department is taking up schemes to ensure that the process of land degradation is slowed down and additional biomass is made available to the people.

Recently the Ministry of Rural Areas & Employment have brought all the area development schemes under the over-arching principle of the micro-watershed as a basic unit for development. The activities taken up would include soil and moisture conservation, afforestation, agro-forestry, horticulture and fodder development. The emphasis would be to fulfil the felt needs of the people both in terms of fuelwood and fodder as also a minor forest produce for economic activities. The project would be implemented through the Zilla Parishads who would be identifying Project Implementation Agencies to do the actual work. A strong training and technology component has been built into the project. In the current year an investment of about Rs. 1000 crores would be made on watershed development through the various schemes of the Ministry.

The Department of Wastelands Development has approved projects to develop about 4 lakh hectares of wastelands in the country. The additional output from these lands would ease the problems of bio-energy in the rural areas which is being discussed today. In addition of the various other suggestions made by me are also taken up in real earnest, perhaps a breakthrough can be possible in this sector.

KEY NOTE ADDRESS

Jayant Patil
Member, Planning Commission
Government of India

The biomass is a traditional source of energy in our country which is used for most of its applications in rural areas. Vast majority of common man still depend on biomass energy for meeting their daily requirement of cooking and heating. It is especially felt that biomass is the only route through which energy can be grown to meet the growing needs of energy demand and to meeting of development needs of the growing population. Biomass the important advantage as a source of energy, it is renewable, it is relatively cheap and adaptable way of storing solar energy and it is generally locally available.

Biomass has been re-organised by Government of India as an important renewable energy resource for meeting the energy needs of the society. Crop residues, agro-industrial residues, fire wood and municipal wastes comprising the biomass resources of the country, contribute significantly in meeting the energy needs especially those of rural areas. The sustainable use of Biomass for energy requirements is considered as an environmental friendly mode of generating energy. Apart from the environmental and economic benefits of biomass, a major benefit lies in its equitable distribution throughout the country. Because of dispersed ownership of this energy source among millions of farmers all over the rural areas, its productive use and value addition will lead to percolation of benefits down to lowest strata of society. Opening up new avenues for income generation in collection, storage and transportation of biomass materials will reduce migration of rural youths to cities. Apart from its rural applications, biomass has been identified for the grid connected power generation programme. Government of India through Ministry of Non-Conventional Energy

Sources has been devoting attention to improving productivity of use of these biomass materials through programmes like improved chullah for rural households, briquetting, biomass gasification for running engines, power generation etc. According to MNES, there is a potential of 17000 MW equivalent Biomass/Bio energy in the country out of which 3500 MW in cogeneration and in addition there is a potential of 1000 MW through urban and industrial wastes out of which only a very small portion could be tapped so far commercial applications.

Bio-Energy Development Programme

The main objective of Bio-Energy Development Programme should be to optimise the productivity of biomass and to develop appropriate technologies for efficient conversion of various forms of Biomass into gaseous, liquid and solid fuels and its utilisation aimed at supplementing the increasing energy demand particularly for thermal mechanical and electrical energy.

Thus, there are three stages involved:

(i) Biomass production/mobilisation;
(ii) Conversion of Biomass into gaseous, solid and liquid fuels; and
(iii) Utilisation of the gaseous, solid and liquid forms of fuels into thermal, mechanical or electrical forms of energy.

(i) Biomass Production/mobilisation

The availability of Biomass can be further classified under five headings:

(a) Biomass grown (plantation);
(b) Agricultural waste—Straw, husk etc.;
(c) Animal waste — Animal dung;
(d) Industial waste — Bagasse, Dairy waste etc.; and
(e) Municipal waste — Garbage, Nightsoil etc.

(ii) Conversion of Biomass into gaseous, solid and liquid forms of fuel

The method of conversion of Biomass into gaseous, liquid and solid forms of fuel, based on available technologies are classified as given below:

(a) Biomass gasification - production of a combustible gas through thermo-chemical process of wood wastes.
(b) Bio methanisation - Production of Methane gas through digestion of Biomass.
(c) Briquetting - For thermal applications.
(d) Extraction/conversion into liquid fuels—Production of alcohol and non-edible vegetable oil to substitute diesel/ petrol.

(iii) Utilisation of gaseous, solid and liquid forms of fuel into thermal mechanical and electrical energy

(a) Biomass gasifier coupled with duel fuel engines for water pumping, running generator sets etc.
(b) Biogas (Methane gas) for heating, lighting and running duel fuel engines.
(c) Biomass in the solid form burnt directly for cooking, heating and industrial applications.
(d) Biomass combustion based power generation.
(e) Liquid fuels like alcohol, non-edible vegetable oil to run engines.
(f) Draught animal energy.

Government of India through Ministry of Non-Conventional Energy Sources, has made good progress in integrating biomass energy into the national economy through various programmes and policies. Development of Bio-Energy Programme was one of the thrust areas during the Eighth Plan period. The National Programme on Biogas and Improved Chullah have already shown significant impact on the rural energy scenario especially for the cooking and heating needs in the rural areas. Other programme based on biomass has been developed and biomass plantation is being promoted. Close coordination and linkage of the biomass research with National Wasteland Development Programme in Social Forestry would be ensured. Biomass generated as industrial and agricultural wastes is converted into useful energy through biomethanisation or through bio-gasification. The bagasse generated as waste in the sugar industry is to be utilised in a large scale for producing power. There are around 420 sugar mills identified for this purpose.

The National Programme on Energy Plantation is being formulated with the objective of producing wood on waste land on economical and sustainable basis and to generate/substitute/replace thermal electrical and mechanical energy involving industrial sector. The research and development activity in the energy plantation programme concentrate on the informance of fast growing fuel wood species and to screening most potential species for plantation under a given set of agro climatic condition.

It is possible to briquette agro-residue and industrial wastes which can be used most efficiently particularly for meeting thermal energy thereby replacing furnace oil and coal. Though briquetting technology has been commercialised in recent years, this technology suffers from various problems and subsequently, this industry has not come on a large scale. This industry, therefore, requires support by way of R&D work to improve efficiency of some of the components of briquetting plant and machineries.

Biomass gasifier system are devices which convert biomass such as wood waste, agricultural and agro-industrial residues etc. to combustible gas through thermochemical process. The gas thus generated could be fed to duel fuel engines or generator sets to produce mechanical or electrical power. Due to indigenous technology development, of biomass gasifier systems of 5 HP and 10 HP for water pumping applications, and 5-100 KW for power generation are commercialy available. IITs at Delhi and Bombay; MKU, Madurai and IISE Bangalore are presently involved in the research and development of biomass gasifier systems for various applications. A number of industries such as saw mills, rice mills, tea/coffee, tobacco processing industries etc. generate a large amount of biomass developing from their own production. Gasifier systems are being promoted in these industries. Ministry of Non-Conventional Energy Sources provides subsidy at the level of 30 per cent for thermal applications and 60 per cent for mechanical and electrical applications upto 100 KW, and for demonstration purposes 100-500 KW.

During 1993-94, MNES has launched a National Programme for cogeneration in sugar industry by providing soft loan or subsidy. A few sugar mills have already been selected for setting up of demonstration projects through MNES subsidy schemes. Other sugar industries are expected to set up cogeneration plants on their own by availing the soft loan facilities being provided by IREDA. The response from the sugar mills so far were not very encouraging. The major constraints are non-availability of capital with industries and lack of proper policies in the states facilitating wheeling of surplus power, banking and buyback of power generated by the cogenerating industries and non-remunerative payment for the power experts after their own consumption of bagasse as fuel in their boilers. In well designed sugar mills it is possible to save bagasse to the tune of 15 per cent of cane. Re-arrangement of power/process steam in the present sugar mills will lead to excess power generation which can be conveniently fed to the state grids.

Human/Animal Energy (HAE) has been recognised as renewable source of energy. The draught animals like bullocks, buffaloes, horses, camels, mules, donkeys etc. play a vital role in our economy by providing essential draught power in agricultural operations, rural/semi urban industries. Animals also offer significant employment and income opportunity to small and marginal farmers. There are about 84 million draught animals in the country which are capable to contributing about 40 million horse power equivalent 30,000 MW of electric power. It is estimated that there are about 15 million animal carts in the country moving freight per year of the order of 25,000 million tonne kilometers. The Eighth Plan strategy and action plan for the Animal Energy Programme include indentification of proven technology, awareness campaign through conferences, exhibitions and publicity through electronic media and involvement of NGOs in promotion and training programme.

Under the National Programme on Improved Chullah, around 196 lakh numbers were set up upto March, 1995. The main objectives of this programme are as given below:

(i) Fuel conservation;
(ii) Elimination/reduction of smoke;
(iii) Environmental upgradation and check on deforestation;
(iv) Reduction in drudgery for woman and children from cooking in smoky kitchen and collection of fuelwood; and
(v) Employment generation.

This programme provides priorities on areas experiencing

deforestation and fuel scarcity, hilly areas, south-eastern areas, semi and slum areas, kitchen of hospitals, hostels etc. and commercial/ industries establishments using wood/other biomass as a fuel. The multi agency approach will continue to be adopted for the implementation of this programme. State nodal departments/ agencies will involve the services of voluntary organisations/ agencies in the non-governmental sector as much as possible.

The installation of improved chulah in the users' households has to be done through self-employed workers. The services of self employed workers will also be utilised for repair and maintenance of old fixed chulhas at the cost of beneficiaries and for promotion of portable chulhas where so preferred by the beneficiary. Portable models are being manufactured in the organised sector and generally used by limited category of users to supplement their cooking requirement or in areas where there is frequent change in venue of cooking. Central subsidy is limited to Rs. 50 per chulah and upto Rs. 75 for SC/ST/Hilly areas is available. In addition, self-employed workers/entrepreneurs are paid Rs. 20 per fixed chulah and Rs. 5 per portable chulah for motivation, local transportation, installation and maintenance for one year.

Under the National Project on Biogas Development; 21.88 lakh biogas plants were so far set up in the country using animal dung as the raw material. R&D on the cost effective design of digestors suitable for various other types of biomass are in progress. Nightsoil based community Biogas plants are also being encouraged. Biogas plants for duel fuel engines for saving diesel are also being promoted. MNES provides subsidy for all these programmes. Central subsidy ranging from Rs. 1500 to Rs. 3500 has been provided to the beneficiaries for setting up of family size biogas plants. Similar to the National Programme on Improved chullah, this programme also provides employment to rural youths for promotion, setting up and maintenance of the plants. MNES provide training facilities and incentives.

DR. O.P. VIMAL MEMORIAL LECTURE

BIO-ENERGY FOR RURAL ENERGISATION

Dr. R.S. Paroda

Introduction

Dr. O.P. Vimal, a doyen in the field of renewable energy, pioneered the development and application of non-conventional energy sources in industrial and rural sectors. A soil chemist by profession, he joined the Department of Non-Conventional Energy Sources as Director (Bio-mass) (now Ministry of Non-Conventional Energy Sources). He visualized the relevance of renewable energy to meet the growing energy needs of the country, particular in the rural sector. Promotion of energy plantation through social forestry for fuelwood and animal feed, improved chullas for cooking, bio-gas for cooking, lighting and application in stationary engines, use of slurry for supplementation of fertilizers, gasification of bio-mass through thermal and chemical conversion into combustible fuels was propagated by him. Besides, he was also concerned with environmental pollution due to excessive use of fossil fuels.

A number of institutions and organisations are promoting the use of renewable energy in the country with Ministry of Non-Conventional Sources as nodal agency. Dr. Vimal appreciated the role of all the agencies and acted as a catalyst in promoting the renewable energy research and development programmes.

Energy Scenario in India

The Energy is the key input for technological, industrial and socio-economic development of a nation. The commercial sources of energy such as coal, fossil fuel, natural gas and with electricity are extensively used, but these resources are finite. It is estimated that 246 million tonnes of coal, 323.5 billion units of electricity, 25.9 million tonnes of diesel oil and 8.7 million tonnes of kerosene are available in the country to meet the energy demand of various sectors. The bulk of POL need of the country is met through importation which affects the economy of the country.

Energy Scenario : (1993-94)

Commercial		
Coal	-	246 million tonnes;
Crude Oil	-	27 million tonnes;
Diesel	-	25.9 million tonnes;
Electricity	-	323.5 billion units
Refinery throughput	-	54.3 million tonnes
Kerosene	-	8.7 million tonnes
Renewable		
Bio-gas	-	1.9 million plants;
Micro hydel	-	110.5 MW
Wind power	-	115 MW

Rural Energy Needs

The energy requirement for the rural and urban sector in India is different due to economic disparity. One of the reasons of poor industrialization of rural sector is due to inadequate or non-availability of assured electricity in the villages. Only about 27% of the total rural dwellings are having electricity for lighting as against 75% in urban dwellings (1991-92). The remaining household use kerosene. This requires about 4 litres of kerosene per month per family. This would require more than 300 million litres of kerosene in rural homes for lighting.

Rural energy needs:

Home management and rural industries	:	66-80%
Agricultural production	:	16-25%
Post-harvest activities	:	2-4%
Animal husbandry and dairying	:	2-5%

The rural energy need can be classified as energy for cottage industries and rural home management and production agriculture and processing. Agro-processing which include flour grinding, dal milling, oil extraction rice milling, dairying, animal feed, plantation crops, Gur and Khandsari etc. are major cottage level industrial units. These units are operated by diesel engines or electric motors. Draught animals are extensively used in sugarcane crushing. It is estimated that about 66-80% of the total energy for the rural sector is used mainly for rural home management and 16-25% for agriculture production. These are met by bio-energy, natural energy, electricity, fossil fuel and coal. Cooking alone consumes 70% of the total energy which is met by wood, crop residues, and dry animal dung. About 3-5 kg of fuel wood per household is required for every day. Yearly requirement of fuel wood for rural sector thus, alone will exceed 154 million tonnes.

The non-commercial sources of energy such as human, animal, and bio-mass are renewable resources of energy. In addition natural sources of energy such as solar, wind and water are available in abundance. The man has exploited all these sources of energy for his need.

Bio-energy source:

Animate power

— Agricultural workers	:	224 million
— Draught animal	:	73 million
Animal dung & excreta	:	1637 million tonnes/year
Bio-mass	:	600 million tonnes
— Crop and crop residues, agro-waste including vegetable, fruits and forestry produce		

Average gas production 160 litre/kg of dry matter

Renewable energy sources have potential to be utilized for generation of electricity for far flung areas, which are otherwise inaccessible to regular energy supply line. The decentralized production of electricity using aero-generators, mini and micro hydel power plants in hilly regions, photovoltaic energy for lifting water, lighting, energisation of household appliances and in communication are now economically feasible. The bio-energy sources are not only pollution free but eco-friendly. In many European countries bio-mass gasification for power generation has become economically feasible, where only a part of agricultural land is required for growing food crops. The economic feasibility of using Biomass in India in power plants, would require in analysis of its alternative uses.

PRODUCTION COST OF ENERGY

Fuel source		Rs./kW	Fuel Source		Rs./kW
Commercial			Renewable		
Coal	:	3.89	Wind	:	3.39
Natural gas	:	2.29	Bio-gas	:	4.23
Diesel	:	5.40	Solar thermal	:	5.60
			Solar photovoltaic	:	25.42
			Tidal	:	10.24

Why Bio-Energy?

Biomass and animate power meets the major energy needs of the rural sector as it is available locally. It is estimated that 35-40% of biomass is utilized for animal feed and the remaining as energy source through direct combustion either for cooking food or for processing of agro-produce. It is estimated that more than 600 million tonnes of biomass is available from various crops residues and agro-wastes. On an average, crop residue is available 2.5 times more than that of the main produce. This potential was visualized by Dr. Vimal and he advocated energy plantation through social forestry programme in unused and waste land in the country. The bio-mass can be utilized for the animal feed as well as for energy. The use of Biomass in place of fossil fuel confers number of in-built advantage such as self reliance, maximum and better use of indigenous resources, minimization of environmental pollution and above all, generation of employment among the rural masses. Besides, plantation reduces depletion of forest, checks erosion of land and also prevent desertification.

The justification for biomass energy:

— Long term energy supply
— Improvement of environment by utilizing wastes
— Better management of surplus agricultural and marginal lands
— Limiting green house effect
— Industrial development and higher rural employment.

Limitations of biomass for power generation:

— Labour intensive to produce, harvest, collect, process and transport, as dispersed in large areas
— High volumes (bulk density 30-180 kg/m^3)
— Economical only in large power plants
— Low reliability
— Lower heat value, cleaning up of gas, NOx emission, dust and other health hazards
— Not convenient in handling decentralized energy plants
— Higher maintenance cost

Limitations of biogas

— Seasonal variation in gas production due to temperature
— Inadequate availability of feed stock
— Slurry handling and disposal
— Poor design construction
— Lack of repair, maintenance and service facilities.

High heat vaue of biomass, 14-18 MJ/kg as against 46 MJ/kg of diesel

Biomass Production for Bio-fuels

Biomass and animate power meet the major energy needs of the rural sector. Fuel wood is estimated to meet between 70-80 per cent of the rural energy budget and 40% of India's overall total energy consumption. The World Bank has estimated fuelwood consumption at 162 million air dried tonnes, a level estimated to be four times the sustained total forest removals for 1992 at 282.4

million m^3, including 10.2 million m^3 of conifer and 270.3 million m^3 for non-conifer material. Ninety per cent of al biomass removal from India's forest is for fuel-wood and energy consumption.

Existing population of 458 million heads of livestock is likely to blow up to nearly 6.50 million by the year 2020, in 1991, the gap between the availability and requirement was 67 and 137 million tonnes respectively in terms of dry and green fodder which would be difficult to meet the forage requirements from 4% of the cultivated area.

To meet the challenge of fuel wood and fodder, concerted efforts are required to augment the productivity of 40% of open forest and rehabilitate the degraded/waste land about 15.23 m ha culturable waste, 36.55 m ha fallow and 20 m ha barren and un-culturable land is available which can be used for energy plantation. As per National Forest Policy (1988), 33% of the plain area must be covered under forest. For achieving this national and global target, ICAR has been working on different agro-forestry interventions in wastelands, arable lands, lands affected by salts, sandunes, ravines etc. and have identified fast growing multi-purpose tree species for different agro-ecological regions of the country.

Besides ICAR and State Agricultural University, Tata Energy Research Institute, DST, CSIR, and many other research organizations under NGOs and general universities have also contributed significantly. Forest department has taken up social forestry to meet partialy the need of fuel and fodder on community land.

Bio-Methanation

Animal dung and excreta (1637 million tonnes/year) can easily be converted into methane gas for cooking, lighting and power generation. This will save nearly 300 million litres of kerosene oil annually. At present only 2.8% of the animal dung is utilized for this. Common weeds like water hyacinth and even the human excreta can be used for biogas for lighting million of houses in the rural area.

Bio-Fuel

The biomass has also potential to be utilized as a supplementary fuel in internal combustion engines by converting into alcohol (bio-fuel). Technology is available for using sugarcane, maize and sorghum for production of alcohol. In Brazil gasohol is a popular fuel being used in transport vehicles. In India, only limited use of alcohol has been made in internal combustion engines. Plant oil is another fuel which has potential to be utilized in I.C. engines.

Indian Agriculture

Although, agriculture sector use only about 16-25% of the total energy used in the rural sector, but the total rural economy is based on it. It contributes 28% of gross domestic products. The food-grains production in the country has increased to all time high of 190 million tonnes. The production of other agricultural commodities has also increased. The gross cropped area has increased to 185.5 mha with a cropping intensity of 1.3. For increased cropping intensity and to achieve higher yield through timely application of inputs such as seeds, fertilizer, chemicals and water, it necessitates higher energy input. Part of the required energy can be met by animate power, but it has to be supplemented by mechanical power. The Indian Council of Agricultural Research has sponsored the All India Coordinated Research Project on Energy Requirement in Agriculture, Utilization of Animal Energy and Renewable Energy, besides other Coordinated Projects to assess the total energy needs for crop production, processing and home management. Crop-wise and region-wise information has been compiled. Projections have been made for the future requirements.

Common uses of agricultural residues and wastes

Following common uses have been observed:

Solid

- Animal feed
- Boards and paper
- Fuel wood, boiler and furnace fuel for heat direct combustion, briquettes and pyrolysed pellets

Liquid

- Chemicals - furfural

— Bio-fuels - alcohol, vegetable oils for I.C. engines

Gaseous

— Gasification - producer gas
— Bio-gas - anaerobic digestion

Bio-fertilizers

For agriculture production there cannot be a better technology of using renewable source of energy than utilization of inexhaustible atmospheric nitrogen through biological nitrogen fixation. Research efforts in ICAR have indicated great promise for rhizobium in legumes and oilseeds, blue green algae and azolla in wetland rice in the temperature range of 20-30° C, azotobacter and azospirillum in cereals, sugarcane etc. These have the potential to fix two to four million tonnes of N into the soil. The Ministry of Agriculture, Government of India, has launched programmes to popularise these and 4183 tonnes of cultures of different strains were produced in 1993-94. It requires a massive programme for production and popularization.

Human Energy

Of the total population of 897 million (1993), 533 million population (62%) is dependent on agriculture. There are a total of 342 million economically active workers of which 224 m (66%) are agricultural workers. These agricultural workers possess equivalent to 15.7 million KW energy. These agricultural workers are engaged in digging of soil, sowing of seeds, transplanting, interculture, harvesting, threshing, cleaning, drying, transport, etc. In the hill agriculture, shifting cultivation, Diara land, vegetable and horticulture cultivation, human power is excessively used. In addition to crop production, the human energy is also utilized in the animal husbandry and fisheries sector. These sectors exclusively depend upon human energy.

Draught Animal Power

Draught animals continue to provide major tractive power for field operations in Indian agriculture. India possessed the finest breeds of draught animals but concerted effort has not been made to upgrade the local progenies. The population of draught animals is estimated to be 73 million. This is equivalent to 18.0 million kW. Draught animal power athough, has potential to generate yearly 27,000 million kWh energy (1500 working hours/year), but its use has been limited to about 10,800 million kWh (600 working hours/year). They are used for seed bed preparation, sowing, interculture, irrigation, threshing and transport. Due to introduction of mechanical power to ensure timeliness in field operation, there has been some reduction in use of draught animals. About 163,000 tractors and 700,000 irrigation pumps are introduced every year. Their population is estimated as 1.5 million tractors and 15 million irrigation pumps. But still more than 55% of land is cultivated by draught animal power.

Social Upliftment and Employment Generation

High animal and human pressures have led to the widespread destruction of native forests and creation of waste land. The main demands of the rural population centered for food, fodder, fuel, fruit, fibre and small timber for agricultural implements and house building. The ICAR piloted a national watershed programme in 1983 with 47 model watersheds spread across the country for overall improvement of the water sheds for stabilizing and maximizing the productivity of food, fodder, fibre, fruit and fuel by linking agriculture with livestock for environment security. The impact at Nartoral (Raipur), Tejpura (Jhansi), Mittemare (Kolar) received National productivity award for excellence. Based on success, encouraged by the gains, Ministry of Agriculture conceived a massive National Watershed Development Project for Rainfed areas in 1990-91. ICAR has further, launched pilot project on technology assessment and refinement through institution village linkage wherein 1000 farm families in cluster village have been adopted at 45 locations to meet the growing demands of different production systems to increase productivity, augment income and improve the quality of life of rural people.

It has been estimated that employment potential exists of 556.5 million man-days for collection, processing, marketing etc of minor forest produce besides, 462.5 million man-days per year for major forest produce. Agro-forestry is one of the few important means for meeting the challenge of unemployment potential exists of 556.5 milion man-days for collection, processing, marketing etc. of minor forest produce besides, 462.5 million man-days per year for major forest produce. Agro-forestry is one of the few important means for meeting the challenge of unemployment. It can take about 30 million people for effective land management through agroforestry-interventions.

Future of Bio-Energy in India

- It would be desirable to pursue an integrated energy management approach to meet the increasing rural energy demands utilizing commercial and renewable energy sources for modernization of agriculture, rural industries and domestic needs.
- The animate power will continue to provide major energy needs of Indian agriculture, especially in small and marginal farms and in hill agriculture. The quality of draught animals however, may have to be improved. The emphasis has so far been, to improve the quality of milch animals.
- Fuel wood for cooking is used in more than 90-95% rural households. Improved chullhas not only conserve energy but also reduce smoke thereby, improve health of women and pollution free environment.
- Bio-gas plants have helped in providing energy for cooking and lighting. Its application in rural industries have not been vigorously pursued as in China. It could not possibly replace electric motor or diesel engines in rural industries. The scientists should look into techno-economic limitation of use of bio-gas for commercial application in rural areas.
- The performance of domestic type Bio-gas plants installed in the country has not been as expected. Low availability of feedstock and poor construction are the possible cause, the engineers should look into it and devise prefabricated design concept of bio-gas plants to improve the quality and reduce the cost of construction.
- Rice and wheat crop residues after combine harvesting and rice husk from rice mills are creating disposal problem in many states. Also about 72 million hectares marginal and

unculturable land is available which has potential to be utilized for energy plantation. Crop residues and wastes which are otherwise not being used or are surplus from cooking and animal feed, need to be utilized through bio-digestion or gasification.

— Use of direct heat energy from bio-gas and producer gas for agro-processing such as Gur making, parboiling of rice, drying of agro-produce has techno-economical feasibility. The quality of processed product is also better.

— The Fuel-Cell using hydrogen from bio-gas and producer gas far stand alone power plant may have prospect in future. Work is already being done in India in few institutions. More efforts are required to establish technical feasibility.

— Increasing cost of POL, would require alternative fuels for mobile power sources. Bio-gas and producer gas can only partially substitute HSD in compression ignition engines. Alcohol and plant oils are compatible with fuel injection system of compression engines. Plant oils have successfully been used in compression ignition engines after transesterification. More research is needed to establish techno-economic viability vis-a-vis other alternative fuels.

Projected energy requirements for the year 2000 for important crops in India

Components of Commercial Energy
(Peta J = 10^{15} J)

	Crop	Total Energy	Commercial	Electricity	Diesel
1.	Paddy	557	342	62	34.8
2.	Maize	37	23	1.4	5.5
3.	Wheat	398	349	33	78
4.	Gram	20	12	2	0.2
5.	Groundnut	123	20	0	0.3
6.	Sugarcane	150	132	3.4	14

(Contd.)

TABLE *(Contd.)*

	Crop	Total Energy	Commercial	Electricity	Diesel
7.	Potatoes	52	40	0	1.2
8.	Sorghum	71	21	1.5	0.3
	Total	1408	939	103	134
Energy in physical units				8740 million kWh	2390 million litre

Source: AICRP on Energy Requirement in Agriculture, ICAR, New Delhi.

PRESIDENTIAL ADDRESS

Dr. Ramendra Singh
President
Bio-Energy Society of India

Bio-Energy Society of India (BESI) is dedicated to the cause of promoting bio-energy research, education and development. The forum created for symposia and workshops organised by BESI render critical review and consideration of latest topics in the field of bio-energy to evolve future strategies. The outcome of these deliberations are published in proceedings and are widely acclaimed as resource material by researchers, planners and administrators. This year's Convention focuses on the 'role of bio-energy in meeting rural energy needs'. We would like to discuss various aspects of bio-energy production and conversion technologies with special reference to social upliftment through rural energisation and employment generation.

The energy requirements of rural areas are also increasing at a fast pace. The decentralised systems of energy generation present tremendous scope to meet these needs. The decentralised systems must, however, be holistic and integrate use of various renewable energy sources. It is a kind of developing matrix system requiring rural appraisal by participatory approach even down to the level of individual farm family. Simultaneous tapping of various energy sources calls for ready availability of user friendly technologies and inputs. Therefore, it is not an easy task. The redeeming aspect of renewable sources, however, is that the inputs are available locally and it is only a question of appropriate technology applications. It is beyond doubt that renewable energy sources or the non-conventional energy sources will continue to dominate the rural energy scene. The advantage of non-conventional energy sources is that they are environmentally safe and present tremendous scope in generation of rural employment avenues.

If we analyse the energy needs of a farm family, it could be broadly divided into two parts i.e., one for the household/home consumption and the other for income generation through farming or farming operations and agro-industrial activities etc. In each of these parts use of bio-energy, particularly the biomass based energy, has been generally found to be cost effective and convenient. Yet the extent to which biomass based energy must become a part of rural households has not been achieved. There are of course problems of technology transfer, application and efficiencies achieved at individual level.

The biomass production and energy conversion systems present great scope for employment generation in rural areas. The advantages accrued are at each level biomass production, energy generation and use. But, then at each level the expertise called for is distinctly unique and entrepreneurial. There is, therefore, an urgent need for human resource development and employment of skilled and semi skilled workers to make significant strides in this area.

Use of wastelands for energy plantations has received an impetus under the new guidelines issued by the Government of India for developing these lands on watershed basis. The scheme envisages integration of work done by different agencies with unified objectivity. What is now required is integration of biomass production with energy conversion system with gainful use of energy in agro-industrial and other spheres of economic development.

Our country produces huge quantities of agro-residues and agro-industrial waste which are poorly utilised at a source of energy due to prblems of handling and logistics. These problems have been solved by briquetting technology. The briquets could be used economically in brick kilns, small furnaces and ovens, gasifiers for thermal/power application and small and baby boilers. Most often end users of briquets find them less cost effective than the conventional fuels such as coal, fuelwood, lignite or oil. At the same time market development of briquet as a product amongst end users has not been done. This requires proper and scientific promotion, comparing advantages of using briquets vis-a-vis conventional fuels.

The biogas programme has helped millions of families in using clean and cost effective energy for cooking. The success of the programme in some states like Gujarat are worth emulating in other potential areas where much is required to be done. The biogas generation from other residues in conjunction or otherwise with dung can also help extend this to those areas where dung is becoming a scarce commodity. The technology of biogas is also matured enough to meet the needs of micro-industrial use. This will help generate employment and income. One example of this is biogas utilization in dairy for providing heat for pasturization.

Biofertilizer is the new field where considerable progress has been made in use of microbes for plant nutrient recycling and generation. These sources are supplementary and complementary to chemical fertilisers. A lot of research and development has been done to make the nutrient generated by biofertilisers cost effective. The farmers are using the biofertilisers where they have found them effective. I feel that more needs to be done to make biofertilisers agronomically effective.

I have highlighted some aspects of biomass in rural energisations.

PART ONE

SOCIAL UPLIFTMENT THROUGH BIO-ENERGY AND ITS USE FOR RURAL ENERGISATION AND EMPLOYMENT

1

GLOBAL BIOMASS PERSPECTIVE

—PRADEEP CHATURVEDI

Biomass energy has traditionally been associated with the energy supply only in the developing world. FAO has observed that around 5% of the world's energy supply is derived from wood (this 5% also is equivalent to about 475 million tonnes of oil). While globally the fuelwood contributes only 5% of primary energy supply, it is an important source of energy in developing countries. It contributes 15% of their total consumption, and in developing Africa (the most dependent region) it averages 58% of consumption. The growth of consumption of wood fuels worldwide has averaged about 2% per year over the past 20 to 30 years, with virtually zero growth in developed countries, but somewhat over 2% in developing countries. The trend in the developing region has been closely related to population growth, but the growth in future years may be expected to be below the rate of growth of the population. As income level rises there is an increasing preference for more convenient alternative fuels. The World Energy Council, in its projection to year 2020 has, however, reflected that till 2020 the energy consumption pattern will be more or less directly proportional to the population increase. The present day technologies will continue to play their important role and it will only be after 2020 that the effect of emerging technologies will be felt and energy consumption growth may delink from population growth rates. By 2020, many current energy end-use device will have been replaced by those just being introduced for commercial use today: new vehicles, new industrial processes and new systems. Many power plants will have been replaced and others will be nearing their end of their useful life. The power horizon to 2050 and beyond means that all energy technologies and devices are likely to have been replaced at least twice.

Biomass Energy Can also Support High Growth Pattern

Energy for Tomorrow's World, the World Energy Council's Report, published in 1993, forecasting energy situation upto 2020 indicates that "energy use (until 2020) will rise and the world will have to rely largely on fossil fuels." No doubt, all further discussion on energy scene was presented at the 16th Congress of the World Energy Council at Tokyo in October last year.

The World Energy Council in its high growth technology intensive scenario (one of the three scenarios) has considered increased role of renewables—especially the biomass. This appears attainable and most likely emerging scenario. The concept of this Chapter is already knit around this hypothesis. 'New renewables' and 'new nuclear' combine to form a 'Bio-Nuc' technology cluster that terms the transmission to a post-fossil age along similar market place dynamics by which fossil fuels phase out traditional energy forms in the course of the 19th century. Natural gas provides the transitional fossil fuel of choice. This strategy leads to a significant degree of decarbonisation of the global energy system. Global carbon emissions are projected to amount to 9 GtC in 2050 compared to 6 GtC today, while primary energy use is projected to increase 25 GtC, which is 3 times higher than at present. It is an illustration of a case in which our rich and clean energy future can overcome some of the challenges of global warming without recourse to stringent environmental policy measures.

This course of application and energy dimension or energy dynamics appears to be much more pragmatic than any of the other cases that the World Energy Council Report has reflected. Development pressure of the developing economies will be the key factor in deciding energy consumption pattern. Some of them being termed as Asian Tigers have already indicated a rapid growth, but this rate of growth is not likely to be sustained over a longer time horizon. Wherever the rates of growth will be higher that will be accompanied by higher technology interventions and energy efficiencies and cleaner environmental technologies. The finance for them will be a constraint because of their own in-built capabilities.

Most of these fastest growing economics have their own sound bases of fossil sources of energy or have developed a strong export market so that they generate sufficient trade surplus to meet energy imports. The story does not end at this point. The concern for the future would largely be for the growth and development of the people from the developing world, i.e., basically from the Sub-Saharan Africa and the South Asia region. Their interest normally does not figure in global reflections. A case of the developing countries, especially of the South Asia and the Sub-Saharan Africa needs specific consideration. They will have a need to enter the high growth rate pattern but will not have sufficient access to fossil sources of energy to meet this scenario of growth and will thus have to depend on largely the technology intensive direction of growth. The economic growth needs and social development needs will have to be segregated. The fossil energy meeting economic growth needs, and renewables meeting the social development needs.

Geo-Politics of Energy supports Larger Role for Biomass

Also at the 16th Congress of the WEC, in his Global Energy Address, Mr Andre Giraud, former Minister of Energy and Industry of France and President and Director General of COGIDEV, France, elaborated on the situation likely to control the growth of energy consumption over the next 25 years. He indicated that in 25 years the geo politics of energy can easily undergo a major transformation in supply of energy in the structure of supply and consumption and the configuration of positions of strengths and weaknesses. The geo-political developments from now to year 2020 will be controlled by three important factors which will interact with each other: technical progress, demographic compulsions of the world populations and their standards of living; and heightened perception of the need to respect the conditions of life on the planet.

Mr. Giraud reflects that it is not inconceivable that the period in question will see the patterns of major new energy sources, in the forefront of which is biomass from solar energy conversion, that is biological fuel conplementing fossil fuels. He said: "I am aware that there is much skepticism here, and that apart from exceptional cases, the cost price of biomass fuels today is 3 times higher than that of traditional fuels. This what was said about nuclear energy less than 5 years before it became competitive. If there is a science which is making rapid progress at this moment, it is Biology: Plant Biology to change the solar cultures which plants are, by technologies to ensure transformation and refining. In my opinion, it is only a matter of time when biomass enters the competitive arena.

The geo-political reflections could be considerable because the areas favourable to biomass production are not the same as those which produce traditional sources of energy. The methods of exploitation are not at all the same either, being most suited to decentralisation and the use of human resources. Consequently, we can anticipate a new sector—agro energy.

With the help of technical progress, we can assume that the energy requirements of humanity will be met overall, with more respect for the environment, and at prices which will not displace much from the situation we know today, leaving aside any political crises which may arise.

Support for Hi Tech Development in Europe for Bio Fuels

A changeover from centralised to decentralised systems will no doubt cause many a problems. But, in countries like India, that practice decentralised functioning, the biomass energy route has the greatest promise.

Two major issues that Mr. Giraud has pointed out are basically relating to 'the environmental necessity' and 'the depleting fossil sources of energy' But he has indicated a very positive shift towards the biomass energy.

Very few people realise that Europe has already entered the area of bio-energy exploration in a very big way. Starting from a modest Export Group Meeting at the FAO Headquarters in September, 1992, Bio-Fuels are now centre-stage. The European Bio-Energy Society which was a low key activity has been energised. From 1993 onwards a number of discussions on incorporation of emergent conversion and combustion technologies for biomass sources have taken place. Large funds have been allocated for intensive and extensive R and D activities under the European Community, with first initiative coming from France. Realising the dimension of R and D support required, the European Union has supported multi-country funded programmes. Dr. Wolfgang Palz, who was once a proponent of solar photo-voltaics, is this year the Chairman of the Organising Committee of the European Biomas Congress starting in Copenhegan on June 24. Such a massive transformation of hearts in people practicing in market economy cannot be just for social cause. It is all guided only by the potential markets opening up in the field of biomass energy.

Major UN Agencies like UNIDO, FAO, UNESCO and UNDP are also orienting their programmes and schemes to support use of biomass as a source of energy. These are a few initial indicators to the fact that biomass is no more a source of energy only for poor in the developing countries. The technological interventions of modern energy conversion process is a major breakthrough due to which bio-fuels are gaining importance in the energy scene. As technologies for conversion of biomass develop and mature the financial projections will also become favourable for utilisation. Initiative of the European Union will certainly be backed up by supporting funds.

Factors Encouraging Shift to Biomass Energy

The most significant factor in the energy scenario over the next 25 years will be the 'shifting populations within a country' and not only of the 'increasing populations'. By this, I mean a shift from rural to urban areas and along with that carrying the habits of the rural masses to the urban centres. The mental blockage of the urbanites towards biomass fuels being a secondary source of energy will be overcome by the rural influx and neo-urbanites. Recent Habitat-II Conference, at Istanbul, from June 3-14, took into consideration the shifting patterns of habitation. It was observed that the problem is getting serious in the developing countries as people move away from rural areas to the urban areas. Cities of developing countries account for almost 50% of the populations and 50% to 80% of Gross National Product. This trend is likely to continue unabated. Overall 80% of the world's urban residents will live in developing countries by 2025

Some of the indicators that have come up need to be referred to much more seriously. The most poor areas of Sub-Saharan Africa and South Asia are expected to see the largest increases in population during the period 1990 to 2020. Sub-Saharan Africa is likely to increase its population from 502 million to 1195 million and the South Asia is expected to increase its population from 1156 million to 1938 million. These increases are of 693 million and 792 million at a growth rate of 24.8% and 28.3%. These dimensions of increases in themselves are substantially detrimental to the growth pattern.

Waste disposal, environmental pollution and energy availability are three major components that have to be seriously considered by the urban planners. These three issues are interrelated. Therefore, future efforts of energy in urban areas need be through the biomass route as far as possible.

There is likely to be a significant expansion of renewables. Traditional uses of renewables and traditional technologies are gradually phased out. Old renewables are gradually replaced by commercially high quality energy carriers, including those from new renewable sources. End use conversion of traditional biomass, however, improves significantly and also has the added benefit of reducing deforestation rate and improving the air quality. Attention of energy laboratories on biomass energy R&D has developed hi-tech and processes that are likely to change the shape of things in the future. The production of biomass route is more favoured and is already promoted in Europe, with efforts to produce liquid fuels.

Production Resources Holds the Key

The land requirement for biomass production for an illustrative high growth rate scenario can be studied further. Depending on productivity increases the land requirement would range between 390 and 690 million hectares by 2020. These additional land areas for both agricultural and biomass growing are in principle available, in particular in developing countries, where most of such land requirements are concentrated. No doubt, they stretch future land requirements to their ultimate limits. Both agricultural and biomass energy are moving into a high tech, high productivity future where eco systems are tightly managed and controlled by man.

I do not wish to say that the land mass be taken out of agriculture and put to additional biomass energy crops. Experiments have shown that multiple cropping systems, where agricultural crops and the energy crops are grown side by side, are already proving financial viable; and the farmers have started adapting them. The contribution of biomass energy grown in a planned manner has a significant role to play at the micro level energy scenario. Examples have been observed in India where Wasteland Development

Schemes in the State of Haryana, after three years of operation, have resulted into stabilising the fuel wood trading price and in some cases even bringing down the price in the local market. This is the case observed, particularly in a fuel wood deficient area, where the prices in the past have always been going up spiral.

Whether it be for population increase or even for the lack of fossil resources, biomass energy is going to play a significant part in the years to come, as the rural populations move on to the urban centres. However, its appropriate exploitation for high quality energy will require appropriate technology development and its disposal in a uniform fashion.

Though the world energy needs will increase substantially, but, at the same time, the energy intensity will improve significantly due to improvement in process technologies, conversion processes and end use devices. Resources availability when looked at the 'final energy' consumption figures will not be a constraint, although from the primary energy point of view it will be a major constraint.

The present trends indicate that the energy transition has the strongest possibility of moving along the third scenario under the high economic growth reflected in the Report of the World Energy Council presented at the 16th Congress at Tokyo. How that can be achieved will require much greater involvement from the society and much greater intervention of the emerging commercial technologies so that maximum benefit is derived by converting biomass sources in the most efficient fashion. Work on the energy efficiency sector will have the largest contribution in promoting biomass sources of energy in the rural and urban areas. This will be the hidden advantage and adaptation of all modern technologies of conversion will really hold the key for making the biomass as Number One Fuel over next 25 years.

2

SOCIAL UPLIFTMENT THROUGH BIO-ENERGY PLANTATION AND EMPLOYMENT GENERATION

—PADMA VASUDEVAN, SATYAWATI SHARMA
AND MIRA MADAN

Social upliftment of the rural sector in India calls for enhancing economic and status simultaneously empowering the people to overcome certain socio-cultural barriers which impede their advancement. . . Thus employment and empowerment go hand in hand and require careful planning to ensure that the benefits actually reach the people. This can be achieved by careful choice of technological system wherein certain inbuilt mechanisms, both in technoeconomic as well as socio-cultural terms, ensure availability, affordability and accessibility.

Activities related to waste land generation is good example as it can create employment specially targeted to the vulnerable section of the society and achieve social upliftment through employment generation besides providing both marketable and locally consumable products satisfying the immediate needs of the poor.

On the national scale both in terms of the availability of waste lands and its distribution in different regions, there is considerable scope for spreading the technologies related to waste land generation involving the poor and creating sustained employment opportunities. In Table 2.1 the total land area of India and waste land distribution are shown. The type of waste land and their distribution are presented in Table 2.2.

Creation of employment opportunities on such waste lands can be by two strategies:

1. Create mandays (md) of work generated for the purpose of

plantation.

2. Through proper choice of plants which can cater to the needs of food, fodder, fuel, fertilizer and other material. The additional employment on the sustainable basis may be created from the biomass grown on these waste lands.

TABLE 2.1: **Land Distribution - India**

1. Total Geographical Area (TGA)	=	326 mha
2. Land Area	=	298 mha
3. Agricultural and Cultivable land (i.e. 45% of TGA)	=	~147 mha
4. Forest Cover (i.e. 22 % of TGA)	=	75 mha
5. Waste Land or Non-Forest Area (i.e. 31.90 of DGA)	=	93.6 mha
6. *Wastelands in Haryana	=	500 thousands ha
7. *Wastelands in District Gurgaon Haryana	=	5,000 ha

TABEL 2.2 : **Estimate of Wastelands in India**

1.	Affected with adverse soil pH or (Salt affected)	7.16 mha
	A. Coastal saline soils	
	(a) Coastal saline soils of arid regions	714th ha
	(b) Deltaic saline soils of the humid regions	1.39 mha
	(c) Acid saline soils	16th ha
	B. Saline soils of black soil region	1.42 mha
	C. Coastal soils of arid and semiarid regions	1.00 mha
	D. Sodic (Alkaline), soils of Indo-gangetic plains	2.50 mha
2.	Wind eroded (Shifting sand dunes, extreme soil moisture stress etc.)	12.9 mha
3.	Water eroded (including varies, water logged)	73.6 mha

Source: Bhumbla & Kharc, 'Estimate of waste lands in India'.

Thus the creation of wages and self employment opportunities will enhance the economic and social status of the poor, increase their confidence level empowering them to participate in discussion and improve environment. By the vary nature of environment working on waste lands will be taken up by the landless, small farmers and women.

To work out a regional and national strategy in this regard the actual generation of employment hours and their distribution over the days and seasons is important. For example a personday (8hrs/day) could involve work done continuously 8 hours or work done over 4 days every day performing 2 hours. Certain types of employment may be home based and flexible timing may be permissible. Also, seasonal variation in employment patterns depend upon the technological systems. The right choice of technology lies in making the system flexible enough to match the constraints faced by the target group. For example, women would need contribute their household chores with flexible hours of working preferably in this. Landless and small farmers may look for auxiliary employment in off seasons.

Keeping this in view careful study is needed on the employment generation pattern through any given technological package. The conceptual framework must be supported by hardcore data. An attempt is made in this regard for finding the employment generation pattern through waste land generation by taking a project for reclamation of waste lands (saline-alkali) in the area of district Gurgaon, Haryana.

STRATEGY

Since planning has to be done for sustained employment over several years the selection of plants for waste land is made both in terms of plants which can survive and stabilize easily on such lands and also subsequently provide employment opportunities to the people in the area. The criteria for selection of the plants is given below:

(1) Salt-tolerant
(2) Hardy and with low inputs
(3) With multiple uses like food, fuel, fibre, fertilizer, fodder medicinal and aromatic, building materials etc.

(4) Mixed plantation to enhance interaction among various soil microbiota for synergistic effect on plant growth, to maintain ecological balance and to enhance income in getting biomass produce all the year round.
In the project plant species selected for plantation on 120 acres of waste lands

(Community or Panchayat lands) are shown in Table 2.3.

TABLE 2.3: **Plants Selected and Planted on 120 Acres of Waste Land**

Name	Uses
Maharookh *Ailanthus excelsa*	Ply, Sh, F, P
Papdi *Pongamia glabra*	M, Ind, P, Fo, Ply
Amaltash *Cassia fistula*	B, D, G, M, of T, Ta, W
Bombay blackwood *Cassia semia*	B, F, G, M, P, T, Ta
Semal *Bauhinia purpurea*	F, G, M, Of, P, T, Ta
Mulberry *Morus alba*	B, F (silk), M, T
(S13, S34, S145, Mandley & local)	
Kikar *Acacia nilotica*	Fe, Fo, G, M, P, T, Ta
Pahadi Papdi *Pongamia pinnata*	M, T, P, Sh, Ply
Bougainvillea *Bougainvillea*	Fe, F, B, M
Gulmohar *Delonix regia*	B, F, M
Australian babool *Acacia auriculiformis*	B, Ec, Fo, G, T, Ta
Subabool *Laucaena Leucocepholea*	B, Ec, Fo, G, Gm, Of, P, T, Ta
Amla *Emblica officinalis*	Fr, M, Ind.
Anar *Punica granatum*	Fr, M, B, Hy
Java Amaltash *Cassia javanica*	B, M, T, Ta, F
Gulabi Amaltash *C. nodusa*	B, D, G, Gm, M, P, T, Ta
Pilkhin *Ficus infectoria*	T, B, Sh, Ec.
Bakain *Melia azadirachta*	B, Fo, G, M, Of, P, T, Ta
Jamun *Schizigium cumini*	B, F, M, Sh
Popular *Populas euphratica*	P, Fo, Ec, M, T, D
Yellow Gold Mohar *Peltophorum ferrugineum*	T, P, M, Gm
Guava *Psidium gaujava*	Fr, M, Ind., Sh.

(Contd.)

TABLE 2.3 *(Contd.)*

Name	Uses
Lasoda *Cordifolia dichotoma*	Fr, M, Ind.
Jungle jalebi *Pithecolobium*	Fr, M, Ec, Gm, Hy
Shisham *Delbergia sisso*	T, F, Ec, Ind.

B - Beautification, D-dye or colouring matter, G - Gum, M - medicinal, Of - fatty oil, T - Timber, Ta - Tannins, P - pulp, Fo - Fodder, F - Fibre, Ply - Plywood, Fe - Fencing, Gm - Green manuring, Sh - Shade, Fr - fruit, Ec - Erosion control, Af - Agroforestry, Ind - Industrial, Hy - Honey, W - Wax.

EMPLOYMENT GENERATED

A. Direct Employment

The direct employment created in terms of persondays depends on the number of plants per unit area depending upon the nature of plants. This may be vary to 4, 9 and 16m^2 amounting to 1000, 440 and 250 plants per acre respectively. Based on plantation experiments a number of mandays of employment generated per acre is shown in Table 2.4.

TABLE 2.4: **Employment Generation through Plantation Programme on Waste Lands**

A. Direct Employment (Through plantation)

1. Area = 1 Acre, Spacing - 4m^2, 9m^2, 16m^2 (Depending upon the variety of plant)

2. Labour Involved:	4m^2 (1000 plants)	9m^2 (440 plants)	16m^2 (250 plants)
(1) Pit making @ 20 pit/m.d.	50 m.d.	20 m.d.	13 m.d.
(2) Plantation and refilling of pit @ 30 plants/m.d.	34 m.d.	15 m.d.	9 m.d.
(3) Aftercare including			
— weeding, watering etc. weeding twice @ 60 plants/m.d.	35 m.d.	16 m.d.	9 m.d.

(Contd.)

TABLE 2.4 *(Contd.)*

	— watering 4 times @ 500 plants/m.d.	8 m.d.	4 m.d.	2 m.d.
	Total labours	127 m.d.	55 m.d.	33 m.d.
3.	Labour involved in subsequent years: (labour Yr/Acre)	45 m.d.	20 m.d.	11 m.d.

B. Auxiliary Employment

1. Nursery Raising

a. Nursery Raising in Nursery beds at Farm Level

(i) Area - 1 acre

(ii) No. of saplings - 40,000

(iii) Labour Involved:

— Nursery bed preparation = 10 m.d.

— Plastic bag filling, seeding etc. = 400 m.d.
@ 100 plastic bag filling/m.d.

— Shifting aftercare for 6 months = 360 m.d.
and for 18 months = 1080 m.d.
@ 2 m.d./day

Total = 770 m.d. (6 months)
1490 m.d. (18 months)

(v) Gross profit from 40,000 saplings = Rs. 1,20,000
@ Rs. 3/Plant (6 months old)
Rs. 8/plant (18 months old) = Rs. 3,20,000

(vi) Expenditure = land rent + Labour + Polythene bags + FYM + Chemicals
= Rs. 2,500 + Rs. 46,200 + Rs. 12000 + Rs. 1000 + Rs. 1000 + Rs. 67,000 (6 months)
= Rs. 7,500 + Rs. 1,10,000 + Rs. 12,000 + Rs. 3,000 + Rs. 2,000 = Rs. 1,35,100 (18 months)

(vi) Net Income = Rs. 1,20,000 – Rs. 67,700 = Rs. 52,300
(6 months)

(Contd.)

TABLE 2.4 *(Contd.)*

Rs. 3,20,000 – Rs. 1,35,100 = Rs. 1,84,900
(18 months)

(b) Angan Nursery

1. Manageable size of nursery raising at household level = 500 saplings/Angan house
2. Labour Involved:
   (i) Bag filling and seeding = 5 m.d.
   (ii) Aftercare
   @ 60 minutes/day = 22.50 m.d.
   upto 6 months
   upto 1½ yrs. = 77.50 m.d.

Total labour = 105.00 m.d.

3. Gross income from 500 saplings = Rs. 1500
   @ Rs. 3/Plant for 6 months old sapling
   Rs. 8/plant for 1½ yr old sapling = Rs. 4,000
4. Net income = Gross income – Expenditure
   Expenditure = Cost of 2 Kg polythene bags + FYM + chemicals
   (Rs. 130 + Rs. 100 + Rs. 100 = Rs. 330)
   Net Income = Rs. 1500 – Rs. 330 = Rs. 1170 + own labour (6 months)
   Rs. 4,000 – Rs. 330 = Rs. 3,670 + own labour (1½ months)

Considering that the district Gurgaon alone has 5000 hectares or 12,500 acres wastelands there is enormous employment potential possibilities. Also. labour input is needed in subsequent years for monitoring the plants for full scale establishment and growth.

It is important to create a forward and backward linkages by creating employment for nursery raising of the plants as well as choosing plants which, on full growth, will be of some use in generating employment continually. In view of this two aspects were considered under the project.

1. Nursery raising
2. Raising multipurpose plants on waste lands so that sericulture can be practiced as a subsidiary occupation

The data on employment creation for nursery raising both on mother nursery of one acre size and Angan nursery raised at household level were implemented. The data are shown in Table 2.4.

B. Indirect Employment

Sericulture is practiced at household level in Haryana and manageable number of eggs is 60 dfls which gives 40 mandays of work with a return of approximately Rs. 2,100.

TABLE 2.5: **Performance and Economics of Sericulture in Haryana**

(i)	Manageable Dfls/household	= 60
(ii)	Mulberry leaves required for 60 dfls	= 600 kg
(iii)	Number of mulberry trees needed	= 300
(iv)	Mulberry leaf yield of 3 yrs old plantation	= 2 kg/plant
(v)	Labour Involved in Silkworm (60 dfls) Rearing (from hatching to cocoon production)	= 40 md
(vi)	Cocoon yield (wet) @ 60 kg/100 dfls	= 36 kg
(vii)	Gross Income @ Rs. 100/kg cocoons	= Rs. 3600
(viii)	Expenditure (excluding labour)	= Rs. 1500
(ix)	Net Profit = Rs. 3600 – Rs. 1500	= Rs. 2,100

Mandays per crop is shown. in Table 2.5. When 4 crops are taken in a year. 160 mandays employment could be generated.

CONCLUSIONS

The project results brought out the potential of employment generation for vulnerable section through waste land reclamation by nursery raising plantation of plants of selected species to waste land establishing the same and using the biomass as the raw materials for entrepreneurship.

It must be emphasized that in this process by mixing the

fuelwood species with others both the bio-energy demand and other needs of poor can be met. It is better to integrate the bio-energy consideration with the biomass production, conversion and utilization system most suitable for the waste land selected and the socio-economic status of the people to be involved in the regeneration.

ACKNOWLEDGEMENT

We gratefully acknowledge the financial support of National Wasteland Development Board (NWDB) for waste land generation acivities, Dept. of Biotechnology (DBT) for nursery raising specially for the creation of Angan nursery and Dept. of Science & Technology (DST) for sericulture programme.

3

RAISING AN ENERGY PLANTATION: Inputs, Techniques and Results

—JOSEPH THOMAS AND R. PERABAGARAN

Technologies to increase the conversion efficiency of coal to electricity are being pursued internationally. This is due to the fact that the calorific value of coal is declining, especially in India. The quantities of coal available are also decreasing and this makes it imperative to utilise it effectively. That increasing energy inputs are required to mine, process and transport the coal is an additional reason for developing these technologies. Integrated gasification combined cycle technology has a high conversion efficiency and is now indigenously available. Bharat Heavy Electricals (BHEL), Trichy has developed the technology and it is ready for commercialisation[1]. The system developed by BHEL can generate 6.2 MWe using coal as a feedstock.

Most of the problems associated with the technology revolve around cleaning the hot gases that exit from the gasifier. Coal, when gasified, gives rise to many substances which need to be removed lest they harm the blades of the gas turbines. Cleaner fuels would perform very well within the existing system with very minor modifications. Dry fuelwood has a calorific value in the range 3500 kcal/kg to 5000 kcal/kg depending on the tree species. It is a much cleaner fuel than coal. Given the declining calorific value of Indian coal it will be the only reasonable substitute in a future without coal.

The question then asked is how much area will it take to provide sufficient fuel wood for an electric power generation plant. A detailed calculation[2] done in 1978 showed that for one MW of generating capacity 1 sqkm would be required. This was based essentially on the biomass output that can be sustainably obtained from a captive energy plantation with the technology available at that time.

The main question then becomes quantifying the annual biomass yield on a unit area basis from a captive plantation. Much of the success of biomass based power generation will depend on what yield levels can be achieved. It is obvious that the yield will depend on the agro-climatic conditions of an area. Yet no effort has been made to systematically establish the best yield for a given agro-climatic zone.

A start in this direction has been made by MCRC and that is reported in this Chapter. MCRC has, in a total area of 60 acres, cultivated various plants and trees. The soil is acidic due to the very intensive agriculture that had been practised on it. The effort was to reclaim the land and quantify which of the plant species could give the best return on investment. Simultaneously, an attempt to raise an energy plantation was also undertaken to try to obtain the best yields. In all cases, organic farming techniques alone have been used, bearing in mind the need to rejuvenate the land.

Inputs and Techniques

The effort undertaken to rejuvenate the land involved planting curryleaf (*Murraya koenigii*), tuberose (*Polianthes tuberosa*), mango (*Mangifera indica*), Neem (*Azadirachta indica*), and Casuarina (*Casuarina equisetifolia*). Of these, casuarina is the only fuel wood species. The others were planted for commercial reasons.

Casuarina was chosen as the fuelwood species over the more popular eucalyptus since it is a fast-growing species and does not evapotranspire as much water as eucalyptus does. Its calorific value is also close to that of Eucalyptus (See Table 3.1).

The output of biomass from a given area will be related to the spacing between plants. To allow for optimal growth of a single plant, the desired space must be available. To increase the yield per unit area, closer spacing can be adopted. This means that harvesting will be done at the end of the exponential growth phase. For casuarina this is between the 4th and 5th years. A spacing of 1m × 1m was adopted for casuarina and a hexagonal planting technique was followed. Hexagonal planting can give an additional 15% of trees compared to square planting.

TABLE 3.2: **Comparison of Calorific Value**[3]

Name of Species	Calorific Value (dry) kcal/kg
Casuarina equisetifolia	4950
Eucalyptus grandis	4900
Eucalyptus tereticornis	4800

Table 3.3 details the inputs to the various species planted, along with the costs. Table 3.2 below gives a summary of the costs and expected returns.

TABLE 3.2: **Inputs and expected returns for different species planted**

Plant	Input costs/acre (Rs.)	Expected returns (Rupees/acre/year)
Curry leaf	28,424	15,000/- (perennial)
Tuberose	18,996	20,000/- (for four years)
Neem	6,400	(not available)
Mango	16,648	30,000 (after 5th year)
Casuarina	15,813	92,000/- (at the end of 5 years)

What can be seen is that the input costs per acre are always in the range of about Rs. 15,000 - 20,000 per acre. The expected returns in rupee terms vary greatly. The primary concern here, however, is with biomass production. At the end of 5 years each casuarina tree is expected to give a pole that weighs about 20 kg on an average at a 36% moisture level. The yield per hectare per year will be 11,500 (trees per hectare) × 20(kg per tree) × 0.64(dry mass fraction)/ 5(years) about 30 tonnes (dry wt.) per hectare per year. This is justified since after one year the entire crop has a fairly uniform height of 12ft to 13 ft and the girth is also uniform. It is to be noted that this plantation has been entirely rainfed. Vettiver (*Vetriveria zizanoides*) has been used in bunds along the fields to conserve moisture and prevent top soil erosion. The mortality rate in the plantation is less than 5%.

TABLE 3.3: **Inputs to the various species planted with costs**

Inputs	Curry leaf	Tuberose	Neem	Mango	Casuarina
Acreage	1.20	1.00	2.50	7.50	15.50
FYM (tonnes)	47.75	30	40	192.50	177
Cost (Rs.)	7637	4800	5600	35141	24840
Tank Silt (tonnes)			32	115	152
Cost (Rs.)			1637	7746	10150
Red Earth (tonnes)	54	24	28	560	
Cost (Rs.)	3600	1600	1400	28114	
Oil Cake (kg)	350	280	177	530	4194
Cost (Rs.)	1330	1070	635	1908	13840
Pesticides (kg)	1				5
Cost (Rs.)	172				115
Seedlings (nos)	3700	9 bags*	300	600	72000
Cost (Rs.)	8250	5105	795	11800	73440
Labour (mandays)**	341	144	51	451	3503
Cost (Rs.)	11270	4741	1675	14894	115600
Ploughing (no. of times)	5	2	2	2	2
Cost (Rs.)	1250	400	1149	3447	7124
Irrigation (diesel lt.)	75	160	112	502	
Cost (Rs.)	600	1280	900	7980	
River sand (tonnes)			12	280	
Cost (Rs.)			600	13830	
Other materials (Rs.) (mud pots, cotton thread, etc.)			1606	1490	

corms: **Weighted average of men and women with respect to wages.

It can be further seen from Table 3.4 that the labour costs are considerable for casuarina. This is a healthy aspect of energy plantations, namely that every MW of power can generate about 56,000 mandays of employment during the planting phase alone. Subsequent operations like digging, watering, de-weeding, etc., will not only promote growth but create more employment. The scope for sustainable power production at a local decentralised level is tremendous.

TABLE 3.4: **Labour inputs and their costs per acre**

	Labour Input Mandays/acre	Cost Rs./acre
Curryleaf	285	9390
Tuberose	144	4741
Neem	20	670
Mango	60	1985
Casuarina	226	7458

Social Impact

The activity undertaken by MCRC has revolved around Kuzhankalcheri village in Chingleput District of Tamil Nadu. Here the villagers are also land owners and cultivate mostly paddy using conventional chemical fertilisers and pesticides. MCRC and its spin-off organisation Chennai Dendro-Tech Private Limited (CDPL) have been the first ones to attempt organic farming in this vegetable and flower production area. CDPL planted paddy on about 10 acres during the first operating season. Using mainly organic manure and natural pesticides, yields a little higher than those in the surrounding areas were achieved. This was an eye-opener to the villagers. A secondary effect was that since a large number of the daily wage workers were from this village a lot of skill transfer took place. The villagers were taught how to plant trees, use organic manure and natural pesticides, etc. The large scale use of neem was quickly adopted by them by the next season. Despite the constant employment guaranteed by the CDPL and MCRC activity in the area, a few disputes have also arisen. These, however, are minor compared to the awareness generated by direct field level intervention.

Summary and Conclusions

The present growth rate of casuarina on this degraded soil shows that, with the right inputs and skills, a high density, highly intensive energy plantation can be successfully nurtured. Yields higher than normal can be obtained using organic farming techniques and natural pesticides. The uniformity of growth seems to be due to the hexagonal planting. A positive impact can also be achieved socially. Decentralised power generation with firewood as a fuel source is a technical reality. In the next four years the best biomass yields obtainable for the area around Kuzhankalcheri village will be established.

Acknowledgement

The work presented here was possible due to financial assistance from PS Apparels, Madras, the Murugappa Group of Companies, Madras and the Department of Science and Technology, Science and Society Division, New Delhi. The authors wish to thank Dr. S. A. Radhakrishnan for his valuable inputs and Mrs. K. Rajalakshmi for her assistance.

NOTES

1. "Power Generation through IGCC", S. Ramanarasu, p. 25, *The Hindu*, Madras November 9, 1995.
2. "Energy Plantations — A case study of the coramandel littoral" C.V. Seshadri, G. Venkataramani and V. Vasanth, Monograph Series on the Engineering of Photosynthetic Systems Vol. 2, MCRC, Madras 600 113, 1978.
3. *Energy from Biomass*, O.P. Vimal and P.D. Tyagi, Agricore Publishing Academy, New Delhi, 1984, pp. 123-125.

PART TWO

BIO-FUEL DEMAND:
Projection and Land Use

4

DEMAND FORECAST FOR BIO-FUELS

—I. NATARAJAN

National Council of Applied Economic Research, New Delhi

Bio-fuels are the dominant source of energy in rural India. In a survey done in India by the National Council of Applied Economic Research, NCAER[1] in 1978-79, it was estimated that bio-fuels accounted for over 90% of the heating energy used by the rural households. The major bio-fuels used are firewood, crop wastes and dung cakes. Firewood alone accounted for more than 55% of the total bio-fuel consumption.

The survey also showed that the rural households consumed over 125 million tonnes of bio-fuels in 1978-79. Of these 67 million tonnes were dung cakes and 30 million tonnes were crop residues. The balance of 79 million tonnes was firewood made up of 20 million tonnes of logs and 59 million tonnes of twigs.

Since 1978-79 the energy demand has grown due to both income and population increase and all the available indications show that the rural households continue to depend heavily on bio-fuels to meet their energy requirements.

During these years there had been an impressive increase in the availability of LPG and kerosene, the two petroleum fuels essentially used in the household sector. These apparently had not made any dent in the pattern of fuel consumption in the rural areas. The Council had conducted two urban surveys[2,3] in the last 10 years to estimate the potential of LPG in urban India. The results of these studies show that the increased availability of kerosene and LPG was mostly absorbed in the urban areas.

Unfortunately, no comprehensive rural energy surveys at the all

India level were done since the 1978-79 survey of the NCAER. There were, of course, a number of regional or small area surveys done by different institutions to understand the pattern of energy consumption in different areas. Most of them are in the nature of case studies and therefore their results do not provide the basis for generalisation.

Nevertheless, various researchers and government bodies had in the past tried to forecast the demand for bio-fuels in general and firewood in particular making use of all the available materials. The range of estimates provided by them was wide. For example, the estimates of firewood demand by the turn of the century varied between 160 million tonnes to over 350 million tonnes. Curiously, a large number of them used the results of the NCAER survey of 1978-79 as the base. Obviously, the methodology and the underlying assumptions were different.

In 1993, the NCAER completed a study[4] to evaluate the "National Programme of Improved Chulha". The findings of this study were based on data collected from a sample of 16,000 households spread over 1,600 villages selected from all the districts in the country. As a part of this study data was collected from the sample households on the energy use and the pattern of fuel consumption. Based on this data an estimate of rural energy consumption and the pattern of fuel use was made and these were presented in a paper[5], to be published shortly. An attempt is made in this Chapter to use the results of these two NCAER studies to provide yet another estimate of demand for bio-fuels in rural India by the turn of the century.

According to the Improved Chulha study, the estimated consumption of energy for heating by the rural households in 1992-93 was 153.8 million tonnes of coal replacement (mtcr). This is about 72% more than what was used in 1978-79. The pattern of fuel consumption for these 2 years is given in Table 4.1.

It is seen that the share of bio-fuels in the 2 years has remained unchanged at about 95%. Firewood consumption has gone up both in quantity and share. An interesting feature is the increase in the use of logs in 1992-93 compared to the earlier year. The survey data also showed that more people collected logs in 1992-93 compared to 1978-79.

TABLE 4.1: **Energy Consumption for Heating in Rural Households**

Fuel	Unit	1978-79			1992-93		
		Qty	CR	Percent	Qty	CR	Percent
Coal/Soft coke	000't	1,143	1,713	1.92	429	584	0.38
Kerosene	000'kl	414	2,279	2.55	1,103	6,831	4.44
Dung cake	000't	66,755	20,087	22.51	86,732	26,151	17.00
Firewood, logs	000't	20,109	16,905	18.95	57,956	49,989	32.49
Firewood, twigs	000't	58,742	31,774	35.62	73,418	44,785	29.11
Crop wastes	000't	29,529	15,531	17.41	24,955	20,530	13.35
Others			913	1.03		4,971	3.23
Total			89202	100.00		153,841	100.00

Forecast for 1999-2000

Both the total heating energy demand and the likely pattern of fuel consumption need to be assessed for the year 1999-2000. It is relatively simple to forecast the total energy demand as its growth is more uniform. The approach adopted in this Chapter is, therefore, to forecast the total heating energy demand and then apportion this between different fuels based on available trends with a few assumptions.

A number of factors affect the level of energy consumption. Two of the most important among these are income and population. To take care of the effect of population increase the following procedure is adopted. The likely average household energy demand is projected for the year 1999-2000. This figure is multiplied with the projected number of households in 1999-2000 to arrive at the demand for total energy in that year.

To estimate the average household energy demand for any future year, the income effect is to be taken into consideration. In order to gauge the effect of a rise in real (household) income on average energy consumption, it is necessary to know the income elasticity of demand for energy.

In the chulha study, data on household income was not collected. The Central Statistical Organisation (CSO) has estimates of personal disposable income at constant prices on a time series. However, these data are not suitable for the present purpose as the income data presented by the CSO is for both urban and rural areas put together whereas for the present exercise data on growth in rural household incomes is required. Moreover, a single elasticity of demand encompassing all levels of income for the entire population neglects the fact that income elasticity of demand for any product varies at different levels of income and also between different segments of the population.

In the absence of data on energy consumption at different levels of income, the following procedure is adopted to forecast the average household heating energy demand for 1999-2000.

Between 1978-79 and 1992-93, the average household heating energy demand grew at 1.9% per annum. This growth may be attributed mostly to the increase in the annual average household income between 1978-79 and 1991-92. It is assumed that is the period between 1992-93 and 1999-2000 also, the average household energy demand would grow at 1.9% per year. This essentially means that the income effect is taken to be the same in the two periods under consideration.

It is likely and there is also evidence for it that the income would grow faster in the coming years compared to the past. But what is relevant is how the income distribution would get affected through this growth. If bulk of this increase accrues to the affluent where the income elasticity for heating energy is likely to be low, the average household energy consumption may not grow faster than what was observed in the past. The results of some of the consumption surveys[6] done by NCAER in the recent past indicate that income in equality has increased in the rural areas. This suggests that the income growth is likely to be relatively more for the affluent.

The demand for average household energy for heating 1999-2000 estimated as above comes to 1550 kgcr.

Population Projections

The rural population of India was estimated at 628.7 million in the 1991 Census. The growth rate between 1981 and 1991 was estimated at 1.8% per year. The census report also reckoned the growth rate for the rural and urban areas put together at 2%. The report of the Expert Committee on Populations[7], estimated the growth rate of the Indian population between 1991 and 2000 under different sets of assumptions giving rise to 3 levels of projections namely low, medium and high. These projections were made prior to the 1991 Census. After the 1991 Census results were released, it was found that the committee's "High" projections were closer to the observed census growth rate. In view of this, it is proposed to use the growth rate under this alternative to estimate the population for 1999-2000. The report adopted an overall growth rate of 1.9% per year between 1991-2000. Rural and urban growth rates were not estimated separately for this alternative. It was therefore decided to adopt the same differential between urban and rural as evidenced between 1981 and 1991 for the years beyond 1991 also. On this basis the average annual growth rate for rural population between 1991-2000 come to 1.7%. The projected rural population in 2000 was estimated at 744 million.

Some of the recent surveys done by the NCAER, indicated the

average household size in the rural areas to be 5.7. Using this average size, the estimated number of rural households in 2000 comes to 130.5 million.

With an average household heating energy demand of 1550 kgcr, the demand for heating energy in rural households in 1999-2000 is estimated at 202.4 mtcr.

Pattern of Fuel Consumption

The next step in the exercise is to apportion this total energy demand between different fuels. It is assumed that the fuels used would be soft coke, kerosene, dung cake, firewood, crop wastes and bio-gas. Commercial fuels like LPG and electricity and other non-conventional energy sources are not considered as their share in the total energy demand is likely to be insignificant.

To assess the likely pattern of fuel use the following precedure is adopted. The demand for all the fuels other than firewood is estimated based on the likely availability and past trend. The residual is attributed to firewood. The method of estimation and underlying assumption are spelt out below.

Coal/Soft Coke

Soft coke is not a preferred fuel in rural India because of its high cost. Between 1978-79 and 1991-92 its consumption had dropped to a third. Its use is mostly in the eastern region near the coal mines where it is collected. This fuel, therefore, is not likely to have a future as a rural energy source. In view of this it is assumed that demand for this fuel in 1999-2000 would be at the same level as in 1992-93 at around 400,000 tonnes.

Kerosene

The use of kerosene for heating in rural areas is limited. Past studies have shown that it is not a preferred fuel for cooking. Its use is mostly confined to water heating for making tea, coffee etc. and bathing. Being a commercial fuel it has not made inroads into rural areas where households invariably obtain their fuel free of cost. Yet, the consumption of kerosene increased from 404,000 kilo litres in 1978-79 to 1103,000 kilo litres in 1991-92, a growth of around 7.5% per annum. Its share in total energy is still small at 4%.

The question is whether this growth can be sustained beyond 1992-93. The answer probably is "No". The use of kerosene as a heating fuel is mostly among the affluent. With the expansion of bio-gas programme, many are likely to switch over to the cleaner bio-gas. Further, with the introduction of parallel marketing of petroleum fuels, the price of kerosene is also likely to go up making this fuel unattractive to the consumer.

In view of this, it is assumed that the demand for kerosene for heating in rural areas would grow at 5% per annum. It may be noted here, that any change in this assumption is not likely to affect the estimated pattern of demand for different fuels as kerosene's share in the energy basket is very small. The estimated demand for kerosene under this assumption comes to 1700 thousand kilo litres.

Bio-gas

There is a growing demand for bio-gas in rural India. From a modest beginning in the early eighties, the bio-gas programme has made rapid strides and it is estimated that upto the end of 1995 around 2.2 million plants were installed in rural India. It is envisaged that this would reach a figure of 3 million by the turn of the century.

In an evaluation study[8] done by NCAER, it was found that around 85% of the plants installed are functional and each of them produce around 400 cu.m. of gas in a year. On the basis of these estimates, the likely availability of bio-gas by 2000 would be 1020 million cu.m.

Dung Cake

Dung cake are an important source of energy in the rural households. Animal dung is also used as manure. The availability of dung is directly related to the cattle population in the country. The rate of growth of animals has been uniform around 1% per annum during the last three decades. It is assumed that the number of animals would continue to grow at the same rate in future also. Further, it is assumed that the proportion of dung used for making dung cakes would not change significantly during the next few years. With these assumptions, the likely use of dung cake in 1999-2000 was estimated using the same average annual growth rate of the animal population.

No allowance was made for the possible diversion of dung to bio-gas plants. For the dung requirements to feed the additional plants (about 1 million) is very small (above 10 million tonnes of wet dung, equivalent of 2 million tonnes of dung cakes). Moreover, it is more likely that the dung used as manure is diverted to bio-gas plants, as these plants in addition to providing a cleaner fuel, also supply enriched manure. The demand for dung cakes estimated as above comes to around 93 million tonnes.

Crop Wastes

Burnable agricultural residues are the second major source of energy widely used by cultivators and farm labourers. The agricultural production witnessed impressive growth during the last 2 decades thanks to the green revolution. But the growth in the yield rates was essentially in foodgrain crops. The by-products of these crops are mostly used for feeding the farm animals. The burnable wastes normally come from crops like cotton, pulses, mesa, oil seeds etc. There has not been any major breakthrough in the yield levels of these crops. The outputs of these crops continue to grow at the historical growth rate. It is assumed that this rate would continue till the turn of the century. The likely availability of crop wastes is therefore projected on the basis of growth rate observed in the past. The estimated demand for crop wastes in 1999-2000 is 38,000 tonnes.

Firewood

The likely contributions of all the other fuels estimated as above is 65.8 mtcr. From the total energy demand of 202.4 mtcr, this quantity is deducted to arrive at a residual of 136.6 mtcr which should be met through firewood.

To convert this energy into physical units of firewood, it is necessary to know the composition of firewood demand in terms of logs and twigs. The twigs burn faster and therefore are less efficient than logs. In other words one kg. of log would give more useful energy than one kg. of twigs.

The NCAER studies observed that the share of logs had increased in the total firewood from 25.6% in 1978-79 to 44.3% in 1992-93. This trend may continue further and it is assumed in 1999-2000 that logs and twigs will have equal share in the total firewood demand. Under this assumption the demand for firewood in 1999-2000 would be around 182 million tonnes.

In the following table, the likely pattern of demand for different fuels estimated as outlined earlier is presented.

Demand for heating energy and likely pattern of fuel use in rural households 1999-2000

Fuel	Unit	Qty	Cr 000't	Percent share
Coal/soft coke	000't	400	600	0.30
Kerosene	Th. k.l.	1,700	10,540	5.21
Bio-gas	Mill.cu.m.	1,020	3,570	1.76
Dung cake	000't	93,000	27,900	13.79
Crop wastes	000't	38,000	23,180	11.46
Firewood	000't	182,133	136,485	67.47
Total			202,275	100.00

Adjustment for Improved Efficiency

The estimates provided so far, have not made any allowance for any likely increase in the efficiency of fuel use in the future. Under the National programme of improved chulhas fuel efficient wood stoves are propagated in the rural areas. Till the end of 1994, around 15 million chulhas were constructed. The last few years saw additions of around 2 million chulhas a year. If this trend continues, by 1999-2000, around 30 million chulhas would have been installed in rural homes.

These chulhas are smokeless and fuel efficient. It is claimed that an improved chulha is three times as efficient as a traditional chulha. If this is true and when all the 30 million chulhas are used in 1999-2000 then a dent would certainly be made in the demand for bio-fuels.

However, recent studies by NCAER show that the improved chulhas installed by the rural households are only 40% more efficient than the traditional chulha. Further hardly 50% of the chulhas are operational. Faulty construction and poor maintenance were cited as the major reasons for the poor performance. The study also noted that around 20% of the existing chulhas become non-functional each year. As a result the new constructions mainly replace the ones that go out of order each year leaving very little for net addition to stock. It is unlikely that this system would improve in the near future. Based on these it is estimated that out of the 30 million chulhas only 9 million would be in use in 1999-2000.

On the basis of a 40% higher efficiency, it is estimated that an improved chulha would save fuels capable of yielding useful energy of around 620 kgcr. For the 9 million chulhas in operation, the total saving would be to the tune of 5.5. mtcr. In terms of firewood it is around 7 million tonnes. Thus the total demand for firewood after allowing for likely improvement in the efficiency of use of fuels would be around 175 million tonnes in 1999-2000.

NOTES

1. Natarajan, I., Domestic Fuel-Survey with Special Reference to Kerosene, NCAER, Feb. 1985.
2. NCAER, Demand for LPG in Urban areas, 1985 (unpublished).
3. NCAER, Demand for LPG in the Northern Region 1995 (unpublished).
4. NCAER, Evaluation Survey of National Programme of Improved Chulha, 1994.
5. Natarajan, I., Trends in Firewood Consumption in Rural India, 1995 (to be Published).
6. NCAER, Market Information Survey of Households (1986, 1987, 1988, 1990, 1992, 1993).
7. Planning Commission, Government of India, Report of Expert Committee on Population Projections, 1987.
8. NCAER, Evaluation Survey of Household Bio-gas Plants set up during Seventh Five Year Plan, 1992 (unpublished).

5

NATIONAL FORESTRY ACTION PLAN (NFAP) FOR INDIA

—A.K. MUKERJI

It is estimated that the lighting of fire was discovered by man some 50,00,000 years ago and since then fuel wood has been used for cooking of food. The Neolithic man started agriculture, fishing and animal husbandry around 10,000 years ago and wood from forests has remained a major source of energy for cooking, preparation of agriculture implements, house building, building of carts and cottage industries, etc. The first evidence of proper management of forests in India, for their multifarious use was found in Kautilya's "Arth Sastra" (321 B.C.) during Mauriya Kingdom when forests were classified based on their management and use as given below:

(a) Reserve forests preserved for hunting pleasure of the king and royalty.
(b) Forests donated to Brahmins and important personages for the use of their family.
(c) Forests for the public use.

Gupta period which followed Maurya's also maintained similar categories for forest management and use upto 673 AD. However, after that the country remained divided into small kingdoms which used forest produce extensively but without much concern about its management. During Moghul period, large areas were created for extension of agriculture though some plantations were raised along the canals, roads, etc.

During British era, initially the forests were overexploited for meeting large supplies of teak and other timber for royal navy, railways, etc. The first step of scientific forest management was initiated with the appointment of Mr. Dietrich Brandis as the first Inspector General of Forests, Government of India in 1864. However, with the rise of population from 238.40 million in 1901 to 361.10 in 1951 and 846.36 in 1991, the pressure on forests specially for fuel increased tremendously leading to over-exploitation and degradation of forest areas and resources.

Fuel Wood Demands and Supplies Scenario

(i) *Demand Scenario:* In the pre-independence era (prior to 1947), there were still substantial forests in the country side on Government, Private and Community lands and the demand was within the manageable proportion as population was only 318.40 million in 1941. The normal practice followed from time immemorial of collection of fuel wood from forest areas for meeting the domestic cooking and cremation needs continued to be practiced unhindered. Adequate supplies were available in and around the village. However, with the rising population, specially in the rural area and development of rural industries requiring fuel wood (brick kiln, jaggery making, tabacco roasting, etc.), the demand of fuel wood increased sharply.

The National Commission of Agriculture (1976) estimated that fuel wood, agricultural wastes and animal dung constituted the major biomass energy source in India accounting for approximately 90% of the energy consumed by households. The major portion of this i.e. about 85% (62% wood fuel, balance 23% agricultural waste and cow dung) was consumed in villages. At national aggregate level, fuel wood and charcoal (from wood) accounted for 46% of total energy consumed in India (National Council for Applied Economic Research-1985). Forest Survey of India (FSI) in its report of 1987 estimated an annual demand of fuel wood at the level of 225 m^3.

The Planning Commission Study Group (1991) on fuel for VIII Plan estimated the demand of fuel wood during 1991, 1996 and 2001 as 306.40, 342.80 and 383.58 million tonnes respectively.

Rural Energy Data Base prepared by Tata Energy Research Institute (TERI) in 1991 indicates the following break-up of bio-fuel requirement.

S.No.	Type of Fuel Source	Consumption Demands (in MT/year)
1.	Fuel wood	252.1
2.	Dung-cake	106.9
3.	Agriculture residue	99.2
		458.2

As such there has been considerable variation in the assessment of the bio-fuel requirement specially for fuel wood in the country. Major portion of the fuel wood is collected free of cost under various rights and concessions granted to the rural and tribal people under the Forest Settlements from the Government owned forests which form more than 90% of the forest area in the country. A policy paper prepared by the Ministry of Environment & Forests in 1992 indicates that this unrecorded removal of fuel wood from forest areas amount to around 220 million tonnes per year in addition to supplies from trees growing in community and private lands.

(ii) *Supplies Situation:* At the time of independence, India had around 40 million ha. of Reserve and Demarcated Protected Forests. An area of nearly 37 million ha. was added to this due to abolition of "Jamindari" system and princely states and subsequently the surrenders made under land Ceiling Act, etc. over a 20 years' period from 1950 to 1970. At the initial stage, due to pressure of population, developmental needs both for forest products and land as well as the uncertainty of ownership and tenure on land and trees, large scale de-forestation took place in this 37 million ha. of forest areas. Moreover, it is estimated by FSI that from 1950 to 1980, nearly 4.3 million ha. of Govt. forest land were also converted to non-forest use e.g. agriculture, horticulture, submergence in the dams, urban and industrial development, etc.

Moreover, the pressure on fuel wood from forest areas increased further due to large scale diversion of private and community land as well as Govt. revenue land which was previously classified as having miscellaneous tree crops to agriculture. The land under agriculture increased from around 134 million ha. in 1951 to 154.70 million ha. in 1991 (nearly 50% of the country's geographical area of 328.8 million ha.)

Thus, 20 million ha. area of forest, tree crop, cultural wastes, etc. got diverted bringing additional pressure on the existing forest land. Moreover, nearly 0.7 million ha. of forest land is being unauthorisedly used for agriculture and 3.5 to 4.00 million ha. of forest area are subjected to shifting cultivation. All these activities have substantially reduced the source of fuel wood supply to rural areas.

FSI in its 1993 assessment through satellites data interpretation have indicated that though legally it is claimed that the Government forest area in India is 76.77 million ha. (23.34% of country's geographical area of 328.8 million ha.), the actual forest cover in India was only 64.01 million ha. (19.47%). Of this only 38.55 million ha. (11.73%) is dense forest cover (crown density 40% and up), 25.03 million ha. (7.61%) of open forests (crown density 10-40%) and 0.42 million ha. (0.13%) mangrove forests. Though India's forest policy lays down that country should have 33% area under forest/tree cover, but FSI has estimated that out of 413 districts only 105 districts have 33% or more of forest cover.

The FSI has also estimated that the growing stock of the forest in the country is around 4740.858 million Cu.M. which works out to rather a low stocking of 74.42 Cu.M. per Ha. of the vegetation cover. Further more, the FSI has estimated that the annual increment put in by this forest growing stock is around 87.62 million Cu.M.

As such, it is clear that withdrawal of fuelwood is far in excess of the production capacity of the present stocking of Indian forests.

Possible Remedial Measures

The National Forest Policy of 1988 lays down that forests should be managed to meet the basic needs of the rural population on a sustainable basis. In view of the serious situation of the growing demand of fuelwood beyond the carrying capacity of our forests, there is an urgent need for adopting remedial measures which are briefly indicated below.

(a) Presently, the investment in Forestry in India is comparatively of a very low level i.e. only 1% of the national development plan amounting to nearly Rs. 8000 million (242.4 million US $) per annum. The estimated withdrawal of biomass (220 million tonnes fuel, 280 million tonnes fodder,

12 Cu.M. timber and few hundred thousand tonnes of gums, resins, medicinal plants, tendu leaves, etc.) priced at US $ 15 per tonne comes to an estimated value of nearly 10 billion US $ per annum.

(b) The current investment can at best lead to raising of a million ha. of plantation in Government areas and another 0.3 to 0.5 million ha. in private and community lands. Taking the average production of 5 metric tonnes of fuel wood per year per ha. from a well-maintained forests, it will require more than 45 million ha. of forest area to be managed only for fuel wood supply. This will be very difficult to ensure specially in view of 14 million ha. of our best forests being preserved as biodiversity reserve under sanctuaries and national parks.

(c) Even considering a rather high productivity of around 8 to 10 tonnes per ha. per annum from well-managed plantation/agroforestry areas, the need of the hour is to raise atleast 20 million ha. of fuel wood plantations in degraded forest areas, community land, etc. during the next 10 years' period to prevent over-exploitation of the natural forests.

(d) Another supportive activity would be to enhance investment in providing more modern cooking equipment, crematorium and animal dung gasification plants in the villages as well as provision of non-conventional energy through wind, solar, etc. sources to relieve the pressure of bio fuels.

(e) Raising of fuel wood trees in community and private lands should be made an essential part of the rural development and employment generation programmes as it is a basic need of rural India. Moreover, 70% of the investment on fuel wood plantation will be in the form of wages thereby creating large scale employment opportunities.

(f) The states having more than 33% area under forest should identify forest deficit districts to bring these upto optimum level of 33% of tree cover through agroforestry and planting in community lands. The states which cannot reach the optimum level of 33% forest or tree cover of their geographical area should work out the optimum level which they can achieve through afforestation and tree planting within and outside forests in the next 20 years period.

(g) Need based afforestation, tree planting, eco-restoration and eco-development programmes should be taken up on the forest and non-forest areas of government, community and private ownership. The attainment of optimum forest or tree cover in the next 20 years' period is being worked out in the State Forest Action Plans now under preparation.

(h) At present the states do not have a Rural Energy Policy and pressure is on the forests for supply of fuelwood for domestic use, headloading for sale in the towns and cities and consumption by rural industries, namely, lime kilns, brick kilns, tobacco curing, etc. At the central level also a clear cut rural energy policy is lacking, it is, therefore, necessary to frame clear policies at the Central and State level.

(i) The Planning Commission should set up an expert group including experts from Ministries of Environment and forests, Non-conventional Energy Sources, Rural Development and Agriculture Departments and NGO's under one of its Members for developing a Rural Energy Policy. This group should clearly bring out the emerging pattern of biomass consumption, i.e., fuelwood, agricultural waste, cow dung, etc. in the rural energy sector and apportion the responsibility of each sector for meeting the demand on sustainable basis. It should then lay down policies and programmes and provide funds accordingly in proportion to the targets fixed for each sector for meeting biomass based energy needs.

(j) The State Governments should also initiate moves for integrating the programme of raising of energy plantations as an essential part of rural development. This sector needs appropriate investment as it will not only meet the essential domestic need but will also create large scale rural employment, conserve soil and water, preserve biodiversity in the natural forests and provide environmental stability to the country.

6

ANALYSIS OF WOOD FUEL SUPPLY FROM NON-FOREST AREAS

—NARAYAN SINGH

Energy requirements of the country have been assessed by the Energy Survey Committee (1965), Fuel Policy Committee (1974), Working Group on Energy Policy (1979) and Energy Demand Screening Group (1986). Most of these have assessed the demand on the basis of energy requirement of 519 Kcal per capita per day. More specifically biofuel requirements have been estimated by the Planning Commission and by Natarajan (1989) and Kaul (1992). Table 6.1 below gives a comparative statement of various estimates.

TABLE 6.1: **Estimated Demand for Fuelwood**

Million tonnes

Year	Natarajan	Kaul	Planning Commission
1980	121.0	—	—
1985	133.3	154.0	—
1989	146.3	—	—
1990	—	165.14	—
1991	—	—	306.4
1995	—	176.28	342.8
2000	163.0	187.42	383.5
2005	—	198.56	—

An analysis of Table 6.1 shows that the estimates vary widely. This becomes all the more confusing if the consumption estimates produced by various agencies are also compared. Table 6.2 below lists three agencies:

TABLE 6.2: Year-1991 **Consumption of Biofuels in Rural Areas**

Million tonnes

Agency	Fuelwood	Animal waste	Agri. waste
NCAER	93.3	83.3	36.7
IREP	169.0	54.2	62.8
TERI	252.1	106.9	99.2

The disparity is because of inadequate sampling. In some cases States have been sampled (NCAER), while in others Agro-climatic zones are taken as units. The size of samples in any case has been inadequate. The latest TERI estimates are based on a survey of 39,000 households in 638 villages spread across 17 States and 14 agroclimatic zones.

During the course of preparation of the National Forestry Action Plan (NFAP) the present supply of fuelwood from forests was estimated at 54.71 million cubic meters and the demand in the year 1996 at 380.44 million cubic meters. The demand will rise to 409.69 in 2001, 439.74 in 2006 and 469.65 in 2011. (1 cubic meter = 0.7 tonnes) These figures are closer to the estimates made by TERI. The forecast of energy demand scenario in 2001 as projected by the Working Group on Energy Policy gives following picture:

With overwhelming dependence of the rural people on fuelwood and vast difference between supply of fuelwood from forests and the demand one is tempted to believe that the adverse supply balance is fulfilled by unregulated felling in forest areas.

Gradual decline in areas under forests and density of crop is no doubt attributable partly to this. However, India's efforts in taking forestry to the people beginning in the Sixth Five Year Plan (1980) and subsequent adoption of a new Policy (1988), led to a massive planting effort both in and outside forest areas. While the first six plans, tree planting targets achieved coverage of less than 8.7 million hectares the achievement in the Seventh Five Year Plan was 8.8 million hectares. The results show that substantial quantities of timber and fuelwood are harvested from non forest areas. A study conducted during the course of preparation of NFAP revealed that as much as 40% of industrial wood is coming to market from land owned by individual farmers.

Two case studies in States as different as Kerala and Haryana, both in terms of physical distance and agro-climatic characteristics are revealing. Before results of the studies are analysed a comparison between the states would be useful.

	Kerala	Haryana
Location	8 to 12 deg N	27 to 31 deg N
Area	3,886,300 Ha	4,421,200 Ha
Population	29 million (1991)	16.5 million (1991)
Forests	940,000 Ha	170,300 Ha
Gr. Stock	99 mill. Cu.M.	1.4 mill. Cu.M.
Climate	Wet	Semi arid

In Kerala homestead farming is typically multi layered and highly productive. A study carried out by the Kerala Forest Research Institute in 1988-89 put the growing stock of homesteads in excess of 112 million cubic meters. It estimates that 92.6% of all timber removals are from non-forest sources. While recorded and unrecorded removals from forests are 740 thousand Cu.M. those from homesteads and estates are estimated to be 13.7 million Cu.M.

In Haryana most of the forests are located close to the Northern boundary with Himachal Pradesh and the official removal of fuelwood from forests is around 0.1 million Cu.M. However, almost all available public lands have been planted with quick growing species. Private plantations have also been raised on a large scale. A study conducted by the National Council of Applied Economic Research (NCAER) in 1978-79 showed that twigs formed bulk of the source of fuelwood in the rural sector. The figures being:

Fuel	Rural	Urban	Total (Mill. Kg)
Solid wood	222.0	79.4	301.4
Twigs	798.7	78.7	877.4
Animal wastes	2,759.0	287.7	3,046.7
Agri Wastes	1,158.4	40.0	1,198.4
Kerosene	8.4	36.9	45.3
LPG	0.4	34.6	35.0

Based on studies made by Forest Survey of India, National Sample Survey, NCAER and other States have been categorised on the basis of consumption of fuelwood. The following table gives average consumption in the States.

(in tonnes)

Firewood	Annual per Capita Consumption	
	Rural	Urban
>0.25	Bihar, Gujarat, Haryana, Maharashtra, Punjab, Tamil Nadu, U.P. West Bengal, Delhi, Chandigarh, Dadra & Nagar Haveli	Andhra Pradesh, Bihar, Gujarat, Haryana, J&K, Karnataka, M.P., Punjab, Rajasthan, Tamil Nadu, U.P., West Bengal, Chandigarh, Delhi, Pondicherry
0.25 to 0.4	Kerala, M.P., Manipur, Pondicherry	Assam, Himachal Pradesh, Kerala, Manipur, Meghalaya, Nagaland, Orissa
0.4 to 0.5	Assam, Nagaland, Orissa, Andaman & Nicobar Islands	Tripura, Andaman & Nicobar Islands
Over 0.5	Arunachal Pradesh, Himachal Pradesh, J&K, Meghalaya, Tripura, Lakshadweep	Arunachal Pradesh, Mizoram

Based on the above and taking total production of States in the table following one comes to the conclusion that as much as 90 per cent of the fuelwood in areas deficient in forests comes from common property resources (CPRs) and individual holdings. It is only in areas near the forests people depend completely on forests for their requirements. This is because fuelwood from forests can be removed free of any levy for domestic use.

Fuelwood supply in States (million M³)

States	Natural Forests		Annual Increment		Plantation	Supply
	Closed	Open	Closed	Open	Increment	Million m³
A.P.	224.80	91.70	2.79	1.14	0.101	4.031
Arunachal	475.99	59.34	5.90	0.74	0.004	6.644
Assam	138.39	35.68	1.72	0.44	0.017	2.177
Bihar	115.11	56.25	1.43	0.70	0.071	2.201
Goa	8.10	1.04	0.10	0.01	0.002	0.112
Gujarat	54.57	22.32	0.68	0.28	0.079	1.039
Haryana	3.00	0.77	0.04	0.01	0.028	0.078
H.P.	81.56	12.31	1.01	0.15	0.023	1.183
J&K	95.68	39.79	1.19	0.49	0.017	1.697
Karnataka	204.00	31.41	2.53	0.39	0.081	3.001
Kerala	71.90	8.03	0.89	0.10	0.045	1.035
M.P.	812.07	167.13	10.07	2.07	0.124	12.264
Maharashtra	225.43	75.57	2.80	0.94	0.076	3.816
Manipur	18.02	51.63	0.60	0.64	0.005	1.245
Meghalaya	27.51	52.26	0.34	0.65	0.005	0.995
Mizoram	38.64	60.63	0.48	0.75	0.020	1.251
Nagaland	31.79	45.54	0.39	0.56	0.013	0.963
Orissa	236.51	83.02	2.93	1.03	0.071	4.031
Punjab	4.39	3.61	0.05	0.04	0.019	0.109
Rajasthan	26.32	39.91	0.33	0.49	0.039	0.859
Sikkim	15.91	3.04	0.20	0.04	0.003	0.243
Tamilnadu	82.71	34.73	1.03	0.43	0.053	1.513
Tripura	16.59	15.59	0.21	0.19	0.008	0.408
U.P.	165.87	46.11	2.06	0.57	0.027	2.657
West Bengal	17.02	11.34	0.21	0.14	0.042	0.392

It has been mentioned earlier that a study in the NFAP process has revealed that as much as 40% of the industrial wood comes out of non forest areas. Logically the same should hold true of fuelwood as about 50% of a tree can be utilised as fuelwood. However, that is not the case because *Prosopis juliflora*, an exotic and long considered a weed has invaded vast tracts of wastelands in semi arid regions growing well even in alkaline and saline areas. It is a good coppicer and thrives under adverse conditions. The fuelwood scenario has been transformed with this species and even the lowly *Lantana camera*, another exotic weed from the same region. *Acacia nilotica*, *Dalbergia Sissoo*, both indigenous species, Eucalyptus, *Cassia siamea, Casurina equisitifolia*, exotics again, have played a significant role in alleviating India's domestic energy problems. A study needs to be undertaken on the contribution of these species towards meeting India's Energy needs.

One factor that is causing serious concern is decline in the areas of CPRs in the villages. Jodha carried out a study of seven States in 89 villages in 1986 and the results show that the area of CPRs declined from 10% to 24% in the villages over 1950-52 to 1882-84 and the average number of persons over every 10 hectares of area increased two and a half times. These areas accounted for fuelwood supplies ranging from 8% in Gujarat to 84% in Andhra Pradesh. Annual income per person per annum from CPRs ranged between Rs. 62 in Gujarat to 770 in Rajasthan. The decline is directly attributable to privatisation of CPRs due to distribution of land among the landless.

Studies in decline of CPRs and impact on fuelwood supplies and the trees planted on lands outside forests need to be studied to chalk out a policy for the future.

7

ANALYSIS OF WOOD ENERGY ISSUES IN NATIONAL LEVEL FIVE YEAR DEVELOPMENT PLANNING EXERCISES

—J.N. MAGGO

India is a large country with a geographical area of nearly 328 million square kilometers. The population of the country was 844 million on March 1, 1991 as per the 1991 Census. The population is likely to cross the 930 million mark by the end of the current year. About 74% of the total population resides in rural areas. The Indian economy continues to be primarily based on agriculture even though the share of agriculture sector in the total GDP has been steadily declining over the years.

2. India became independent in 1947. The planning process was initiated in 1951 with the launching of the First Five Year Plan. Since then eight Five Yer Plans have been formulated and executed. There were a number of Annual Plans also during this period. The work on formulation of the Ninth Plan is to be initiated shortly.

3. Energy is one of the key inputs in the developing of the country's economy. Energy is consumed in a variety of forms. Fuelwood, animal waste and agricultural residue constitute the traditional sources of energy that continue to meet the bulk of the energy requirement, particularly in the rural households. These traditional or the "noncommercial" fuels are gradually getting replaced by the commercial fuels like electricity, coal, lignite, petroleum products and natural gas. The share of noncommercial fuels in the total primary energy supply has declined from more than 70% in 1953 to less than 50% in 1993. The consumption of noncommercial fuels in 1993-94 is estimated to be 118.8 MTOE (million tonnes of oil equivalent). Fuelwood comprises the major

share and accounts for a little more than 65% of this consumption. In actual units the fuelwood consumption is estimated to be around 160 million tonnes annually.

4. The Planning Commission is responsible for preparing both the Five Year and Annual Plans for the various sectors of the economy. The overall plan perspective is provided by the Perspective Planning Division while inputs for the sectoral plans are provided by the concerned Divisions. These inputs are based on an eiaborate process of consultations with the various Central Ministries/Departments and undertakings concerned with the sector. At the time of formulation of a Five Year Plan the Planning Commission sets up a number of Working Groups. There is a working group for each of the sub-sectors of energy. The working groups comprise of representatives of the concerned major producing industries, the consuming sectors as well as of the Planning Commission and the administrative Ministries. The sectoral estimates of energy demand provided by the working groups are then used by the Perspective Planning Division of the Planning Commission to work out the sectoral demand estimates consistent with the growth profile of the economy. There are other models on energy demand and supply optimisation which have been developed within the Planning Commission as well as by other agencies like the Tata Energy Research Institute.

5. The subject of energy sector development has been receiving the due consideration of the Government in view of its importance for sustaining the growth of the economy and meeting the basic needs of the people at large. Integrated energy planning was recognised as an essential element of development planning as early as in the 60's. A number of high level Committees/Working Groups were set up in the past to examine in detail the various issues and make recommendations concerning the development of this sector in the short, medium and the long term. The recommendations made by these Committees/Working Groups came in handy while formulating various Five Year Plans. The Energy Survey of India Committee was constituted in 1963 to study, "the present and perspective demands and supplies of energy, both total and in respect of constituents of energy on a national, regional and sectoral basis". The report of the Energy Survey of India Committee was submitted in 1965 well before the Fourth Plan was finalised. Subsequently, the Report of the Fuel Policy Committee was prepared at the time of formulation of the Fifth

Plan; Report of the Working Group on Energy Policy at the time of the Sixth Plan and recommendations made by the Advisory Board on Energy at the time of the Seventh Plan formulation. The Planning Commission has recently constituted the Energy Policy Committee. The recommendations made by this committee will be kept in view while finalising the Ninth Five Year Plan.

6. As has been mentioned above, estimation of energy demand for different fuels by the consuming sectors form the initial step in the process of energy planning. In the case of non-commercial fuels, including fuelwood, the demand mostly arises from the household sector for meeting the cooking needs of the households. The data base for energy consumption in the household sector especially for the non-commercial fuels is not comprehensive. This is because non-commercial fuels are mostly gathered outside the market place and the transactions are rarely recorded. Primary data are available from field surveys conducted by the National Sample Survey Organisation (NSSO), National Council of Applied Economic Research (NCAER) and other research institutes and individuals. However, these surveys are far too scattered to give comprehensive picture of the regional variations in the pattern of energy consumption of these fuels.

7. Household sector is the major consumer of energy. Non-commercial sources of energy play a major role in the energy requirements of the household sector, particularly in the rural areas. The methodology of estimation of household energy demand was first given by the Energy Survey of India (ESIC) in 1965. Based on the results of the sample surveys carried out by the agencies mentioned above, the ESIC adopted a norm of 0.39 tonnes of coal replacement (tcr) of energy consumption requirement per person per year in the urban households and 0.38 tonne of coal replacement of per capita energy consumption per annum in the rural areas. After so determining the total energy requirement in the household sector, the projected contribution of commercial sources of energy was subtracted from it to arrive at the share of non-commercial fuels. The same was then apportioned among of fuelwood, agricultural waste and cowdung in the proportion of 65:20:15. This is only to indicate that the available estimates of non-commercial energy consumption in the household sector yield only rough estimates rather than precise estimates. This methodology continued to be adopted till the mid-80's The Advisory Board on Energy adopted the concept of

useful energy requirement per person in the household sector. In the report on "A Perspective on Energy Demand and Supply in India in 2004-05 published in 1985, the ABE adopted a norm of 650 Kilocalories of useful energy requirement per person per day for cooking and space heating in the year 2004-05. The fuel mix for the year under consideration was then worked out on the basis of population, the rural-urban distribution of population etc. The fuel mix adopted for the exercise was as per the Domestic Fuel Survey of the NCAER for the year 1978-79. Later in 1986 the Energy Demand Screening Group (EDSG) set up by the Planning Commission modified the consumption requirement of useful energy to 520 kilocalories per person per day in the rural areas. Other studies carried out later like the Sectoral Energy Demand in India also estimated the useful energy requirement in 2009-10 as 641 kilocalories per person per day.

8. The requirement of fuelwood as estimated from the exercises was of the order of 200 million tonnes per annum. This is an indicator of the importance of fuelwood as an energy source. Fuelwood is obtained not only from forests but also from trees which exist outside the forests. However, the area under forests is no indicator of the quantity of fuelwood available in the forests which could better be described by an index called "Growing Stock" and the annual increment in growth of the forests called "Mean Annual Increment". These indices would depend on the nature of the forest, the varieties of trees, etc. Adequate information was not available about the yields each of the tree species would provide. Growing stock estimates made by different agencies gave different results. The Pre-investment Survey, carried out in 1965 in collaboration with the FAO, investigated the forest resources in three small selected areas of the country. The results of the survey indicated that the quantity of growing stock would be widely differing in the different areas. Furthermore, the estimates of growing stocks made were with reference to timber yields only and excluded the yields from bark, twigs and branches. The Gross Annual Increment of forest represent the additional wood produced each year in the forests. If the extent of removal of wood from the forests is at the same level as the annual increment and if such extraction is scientifically managed, the stock of wood in the forest would remain unchanged. In such cases where the removal is more than the increment, problems will arise. Keeping all this in view, the Working Group on Energy Policy in its report in

1979 opined that state of knowledge about the fuelwood resources was very inadequate. The Group recommended that the methodology for assessment of fuelwood resources be examined by a committee of technical experts including experts on rural energy and fuel technology.

9. The Fuelwood Study Committee was constituted by the Planning Commission in 1981 to study all aspects of fuelwood including production, consumption and utilisation of by-products and other related matters for energy purposes, in the light of past experience in India and other countries and to work out a suitable proposal for inclusion in the plan. The Fuelwood Study Committee examined the subject in detail and gave a set of recommendations in this regard. In view of the major role fuelwood plays in meeting the energy needs of the rural and urban populace, augmenting fuelwood supplies is of prime importance and calls for various additional measures to be initiated and implemented effectively. Keeping in view the data on energy consumption pattern in the rural areas and availability of commercial and non-commercial fuels, the committee stressed the need for a dynamic integrated and long term action plan for promoting the growth of fuelwood plantations in the country.

10. The recommendations made by the Fuelwood Study Committee included carrying out a survey to assess the availability and production of non-commercial fuels; identification of forest areas for intensive rural fuelwood production and identification of lands of different categories available and suitable for fuelwood and/ or biomass production in each State. It recommended introduction of improved management programmes for increasing the fuelwood availability from lands of different categories identified above. The Committee further stressed the need for taking up a strong and time bound R&D programme in the area as well complementary programme for increasing the efficiency of the devices using wood and wood products and development of other energy sources like biogas plants. The Committee further recommended identifying fuelwood as one of the basic needs for coverage under the minimum needs programme.

11. The Sixth Five year Plan (1980-85) recognised firewood as the most important non-commercial fuel which was becoming increasingly scarce and also getting commercialised. Energy forestry was also recognised as an important component of the energy strategy as it was felt that the rural communities will continue to

depend on firewood for several decades to come. A large programme of fuel and farm forestry was taken up in the Plan. Also, an Integrated Rural Energy Programme (IREP) was initiated in the Sixth Plan on a pilot scale with an objective of providing a mix of energy options for meeting the energy needs of the rural areas on an area-wise basis in the most cost-effective manner. To implement this programme, the Planning Commission set up a separate Rural Energy Division. A Rural Energy Planning Exercise was commissioned by Planning Commission to develop the design and approach of planning and implementing area-based block-level integrated rural energy projects.

12. The Advisory Board on Energy brought out a report "Towards A Perspective on Energy Demand and Supply in India in 2004-05" in 1985. The main conclusion of the report pertaining to supply options, analysed Statewise, in the case of firewood were that at least 60 million hectares of land (comprising open/degraded forest land, barren and unculturable land, culturable wasteland, land other than current fallows and farm perimeter land) might be available for raising fuelwood/fodder plantations. Given the acute pressure on land and the other uses it could be put to, not more than half of the land area thus identified could actually be used for plantation purposes. The Study brought out that the productivity of the Indian forest lands is quite low and need to be increased. On the basis of the yields expected, present day chullah efficiencies and availability of land, it may be difficult to meet the fuelwood requirements of certain areas of the country.

13. As has been mentioned above, India consumes nearly 119 million tonnes of oil equivalent (MTOE) of non-commercial energy presently. According to the report Sectoral Energy Demand in India (1991), despite large substitution by the Commercial energy, the level of non-commercial fuels will still be at a level of 103 MTOE in 2004 and 84 MTOE in the year 2009. As it is to be expected, the bulk of this will have to come from fuelwood. Even if afforestation were to be taken up on a large scale during the next 15 to 20 years, it would still leave a sizeable deficit of forest cover in terms of what is optimally required and what is actually available. In other words, the emerging scenario of energy supply will imply continued heavy dependence on fuelwood of energy during the 20 years to come and, consequently, continued "mining" of the tree cover with the accompanying ecological degradation. In the case of both crop

residue and dung, since the competing demands for these forms of biomass would also increase simultaneously, it is difficult to visualise any incremental availability of either dung or crop residue without its further adverse impact on soil fertility and livestock productivity. It is important to note here that if commercial energy supplies fail to materialise to the extent envisaged in future, there will be a heavier burden imposed on the traditional fuels leading to an aggravation of the problems referred to above. The present balance is a fragile one. Added to this, commercial energy supplies (oil) are subject to the vagaries of the external markets.

14. Another aspect that needs to be considered is that the inefficient burning of biomass emits smoke causing health hazard to beneficiaries who are largely in the lower income groups of the rural and urban areas. In the normal course, the major objective of energy planning should be either to replace these traditional fuels completely or introduce cleaner and more efficient technologies for the conversion of biomass into useful energy for cooking. The latter option will call for major investments as well as technological upgradation. In the case of dung, biogas is a well established technology. If the traditional fuels are to be replaced completely, it will require wither kerosene or coal-based fuels. Taking into account the relative calorific values as well as the end-use efficiencies, the requirement of equivalent quantities of coal and kerosene for replacing these fuels have been estimated. In other words, to protect the rural and the urban poor from exposure to excessive emission of smoke, it is necessary to step up fossil fuel supplies, either petroleum products or coal or both.

15. Even from the point of view of the long term effect of carbon dioxide emission on the environment, it should be noted that one tonne of coal used in the household sector will save 2.25 coal-equivalent tonnes of fuelwood as the end-use efficiencies of the two fuels are in the ratio of 2.25:1. Similarly, one tonne of kerosene used in this sector will save 5.6 kerosene equivalent tonnes of fuelwood. This is a typical example of how fossil fuel consumption at the margin can be environmentally benign in developing countries like India.

16. While such transition from traditional to commercial fuels will, therefore, be welcome from the point of view of local as well as global environmental concerns, it is beset with other problems such as affordability of the low income groups who will then be required

to use the more costly commercial fuels in the place of non-commercial fuels. According to one study, the expenditure of household energy consumption in the lower income groups in both rural and urban areas amounts to 10 to 14% of the total household expenditure on all items. Increased dependence on commercial energy will, therefore, impose a heavier burden on the poor. The only way to resolve this problem is to enhance the purchasing power of the low income groups of the community.

REFERENCES

1. Report of the Energy Survey of India Committee, 1965, Government of India.
2. Report of the Fuel Policy Committee, 1974, Government of India.
3. Report of the Working Group on Energy Policy, 1979, Planning Commission, Government of India.
4. Report of the Fuelwood Study Committee, 1982, Planning Commission, Government of India.
5. Towards A Perspective of Energy Demand and Supply in India in 2004-05, Advisory Board on Energy, 1985, Government of India.
6. Report of the Energy Demand Screening Group, 1986, Planning Commission, Government of India.
7. Sectoral Energy Demand in India, 1991. Study carried out in Planning Commission in Cooperation with ESCAP, UNDP and Government of France. The report has been published by ESCAP.
8. Review of the Energy Planning Practices and Methodologies in India, 1992. Study prepared by E.A.S. Sarma, J.N. Maggo and A.S. Sachdeva for Asia and Pacific Development Centre, Kuala Lumpur, Malaysia.
9. Sixth Five Year Plan, Planning Commission, Government of India.
10. Seventh Five Year Plan, Planning Commission, Government of India.
11. E.A.S. Sarma, J.N. Maggo and A.S. Sachdeva, Energy, Environment and Quality of Life, published in *Science and Quality of Life*, Ed. Dr. S.Z. Qasim, pp. 105-122, The offsetters.
12. R. Bhatia and R. Sharma, 1986, Energy Pricing Strategies, Vol. II, India, Meeting Basic Energy Needs of the Poor, REDP, UNDP/ESCAP/ILO/ Asian Employment Programme (ARTEP), Regional Office for Asia and Pacific, New Delhi.

PART THREE

FINANCING OF BIO-FUEL PROJECTS

8

CO-GENERATION IN SUGAR INDUSTRIES
Potential and Prospects in India

—B.S.K. NAIDU, K. BALAKRISHNA REDDY AND NISHA MENON

India's involvement in Renewable Energy movement is being reported to be the second most extensive in the developing countries next only to China. The growth in this sector can be attributed mainly to the favourable policy and financing environment existing in the country and concern for environment conservation and sustainable development. Several international development financial institutions like the World Bank (WB). Asian Development Bank (ADB) and the United States Agency for International Development (USAID) have shown interest and are involved in the financing of renewable energy projects.

In order to meet the growing demand for power supply, the Government of India is likely to add some 20,000 MW generating capacity through conventional sources besides 2000 MW through renewables including co-generation during the VIII five-year plan (1992-97).

India is currently the largest producer of cane sugar in the world, accounting for 10% of the world production. Sugar is a growing industry with the cane area, yield and recovery of sugar increasing over the decades, though there are cyclical variations from year to year. Though the concept of bagasse based co-generation has always been practiced by sugar mills, there has of late been growing awareness in the sugar industry of the advantages of installing 'high efficiency' bagasse based co-generation systems. The sugar mills have been further encouraged by favourable policies announced by the Government of India, the State Governments and various State Electricity Boards.

Co-Generation—The Concept

- Cogeneration is defined as the concurrent generation of process heat and motive power in an industry by the sequential use of energy from a common fuel source.
- Depending on the quality of process heat required, co-generation may be based on topping cycle or on bottoming cycle.
- In bottoming cycle systems, heat is required for the process at high temperature and hence power is generated through a suitable Waste Heat Recovery System.
- On the other hand, in the topping cycle system, heat is required for the process at low temperatures and therefore power generation is taken up first. All sugar mills employ this cycle for co-generating power and heat.

COGENERATION

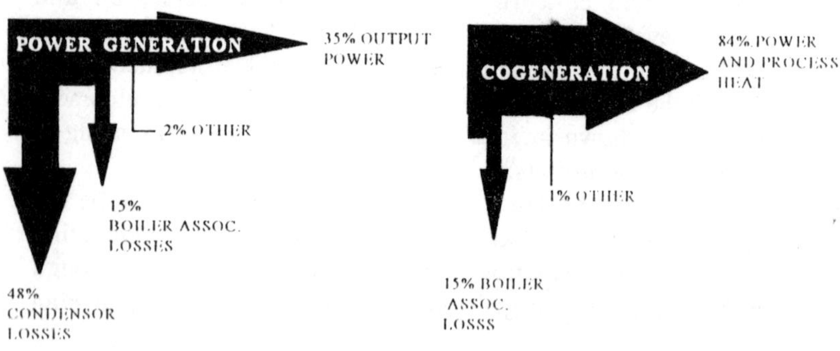

The Process of manufacturing crystal sugar requires steam. In existing co-generation systems steam is generated in low pressure boilers by using bagasse-the woody fibrous residue of crushed cane, as fuel. This system was developed when possibility of exporting power to the grid was not envisaged. Further, since the storage of large quantities of combustible bagasse in the premises of the sugar mill was not an advisable option, most of the boilers were designed so as to use almost the entire quantity of bagasse produced. There is thus a 'built-in energy inefficiency' in the sugar factories using all the bagasse as fuel for their low pressure boilers. By upgrading the steam parameters, the sugar industry can produce electricity far in excess of their own requirement and sell the surplus power to the grid.

Benefits of Adopting High-Efficiency Co-Generation Systems in Sugar Mills

1. Not depending on external power at all, sugar plants can be located nearer the sugar growing areas, thereby saving on transportation cost of sugar cane.
2. An efficient and sustained co-generation enables the plant to isolate itself from the vagaries of power.
3. Power generation using bagasse is environmentally cleaner as bagasse produces very little fly ash, and no sulphur.
4. Net contribution to greenhouse effect from the bagasse based co-generating plant is zero, since the carbon-di-oxide absorbed by the sugar cane grown in more than the one emitted by the co-generating plant.
5. Low gestation (18-24 months) as compared to 96-120 months for a coal based power plant.
6. Low capital investment. Recurring costs are also lesser compared to fossil fuel based power plants.
7. Use of a totally renewable source of energy. Total saving in the mining, extraction and long distance transportation expenses of fossil fuels.
8. Rural location of sugar mills enables co-generated power to be directly fed to the local sub station, consequently minimizing T&D losses and the requirement of long feeder lines.
9. Benefits of quick return on cogeneration capital investment.
10. Saves the expenditure on safe storage and disposal of bagasse.
11. A cogeneration plant places no financial or administrative burden on the utility as it is executed and managed by the sugar factory.
12. Power is generated at a lower cost in cogenerating systems and pay back periods are shorter.
13. Provides an initiative to sugar mills to concentrate more on conservation of energy and reduction of steam consumption

thereby improving their profitability of operation.

14. Surplus power generation in sugar factory is ideally suited for rural electrification and for energising irrigation pumps and industrial and agro-based units in the villages.
15. Even among the renewable energy projects, this perhaps is the only sector which has readily available infrastructure.

Potential for Bagasse Based Co-generation: The Indian Sugar Industry

Sugar production in India takes place in 18 States. Table 8.1 gives the statewise break-up of the 427 Indian sugar mills in different states.

TABLE 8.1: **Statewise Sectoral Break-up of Sugar Mills in India**

Sl. No.	State	Joint Sector	Public Sector	Co-op. Sector	Total
01.	Uttar Pradesh	44	35	31	110
02.	Maharashtra	5	-	104	109
03.	Andhra Pradesh	10	8	18	36
04.	Tamil Nadu	14	3	15	32
05.	Karnataka	11	3	17	31
06.	Bihar	14	15	-	29
07.	Gujarat	-	-	19	19
08.	Punjab	4	-	15	19
09.	Haryana	1	-	10	11
10.	Orissa	7	-	1	8
11.	Madhya Pradesh	2	3	3	8
12.	Kerala	1	-	2	3
13.	Rajasthan	1	1	1	3
14.	Assam	1	1	1	3
15.	West Bengal	1	1	-	2
16.	Pondicherry	1	-	1	2
17.	Nagaland	-	1	-	1
18.	Goa	-	-	1	1
	Total	117	71	239	427

The cane crushing season in India is during the period of 'November to May' and lasts on an average for 200 days.

The size of sugar plants in India, in terms of daily cane crushing capacity, varies from 1250 TCD to 5000 TCD and above. The minimum size of a sugar mill has now been stipulated at 2500 TCD.

Table 8.2 gives the state-wise break-up of sugar mills in India based on their licensed cane crushing capacity.

TABLE 8.2: **Capacity-wise Break-up of Sugar Mills in India**

Sl. No.	State	≤ 1250 TCD	< 2500 TCD	2500 TCD	< 5000 TCD	> 5000 TCD	Total
01.	Uttar Pradesh	44	18	29	12	7	110
02.	Maharashtra	44	18	26	15	6	109
03.	Andhra Pradesh	22	4	6	1	3	36
04.	Tamil Nadu	6	7	13	4	2	32
05.	Karnataka	15	5	6	1	4	31
06.	Bihar	15	10	3	0	1	29
07.	Gujarat	7	1	4	3	4	19
08.	Punjab	6	2	9	1	1	19
09.	Haryana	5	2	3	0	1	11
10.	Orissa	5	0	3	0	0	8
11.	Madhya Pradesh	7	1	0	0	0	8
12.	Kerala	3	0	0	0	0	3
13.	Rajasthan	2	1	0	0	0	3
14.	Assam	3	0	0	0	0	3
15.	West Bengal	2	0	0	0	0	2
16.	Pondicherry	1	1	0	0	0	2
17.	Nagaland	1	0	0	0	0	1
18.	Goa	1	0	0	0	0	1
	Total	189	70	102	37	29	427

TCD: Tonnes of cane crushed per day.

A typical sugar mill has the following features:

1. Bagasse produced by crushing cane is 30-35% of the cane crushed.
2. Bagasse is burnt to generate steam in boilers. The steam pressures encountered range from 15 in old plants, 21 in plants installed in the seventies and 32 to 43 kg/cm^2 in mills installed since the mid-eighties.
3. Operating at low pressures, the mill can use almost the entire bagasse for process steam generation thereby avoiding accumulation of bagasse and eliminating the need to make provisions to store bagasse.
4. Process steam consumption is 50-55% of the cane crushed.
5. Mill wet bagasse produced contains about 50% moisture-(48-52%). The gross calorific value of mill wet bagasse (50% moisture) is 2280 kCal/kg. The proximate and ultimate analysis of mill wet bagasse is as under:

Proximate Analysis	Ultimate Analysis
Fixed Carbon - 11.5%	Carbon - 22.5%
Volatiles - 37%	Hydrogen - 3.0%
Moisture - 50%	Oxygen - 23.0%
Ash - 1.5%	Ash - 1.5%
	Sulphur - traces

Co-Generation Potential in India

A number of studies have been conducted to estimate the potential for bagasse based co-generation. The results of some of the more recent studies are as under:

1. In 1993, an MNES task force formed to study the potential of bagasse based co-generation, have estimated the potential at 3500 MW.
2. On the other hand, the Tata Energy Research Institute have estimated India's potential for co-generation in sugar mills as 5100 MW.
3. A third study conducted by USAID/Winrock International/ IDEA Inc., in May, 1993 have reported that efficient

co-generation could yield an additional 3800 MW of electric power.

4. The most recent estimation was made by an IFREE/dLA/IDS team, who have estimated the potential as 2800 MW with highly efficient mill operations and high pressure conditions.

However, bagasse based co-generation has a major drawback that the annual production period is usually limited to six to seven months. This period does not normally match with the seasonal variation in electric load demand. Bagasse availability period coincides with the surplus hydel peaks in some states, while in the peak summer months, sugar factories are shut down due to non-availability of cane. This constraint may be circumvented by spreading electric generation throughout the year, using available bagasse or by resorting to other fuels in the off-season.

National Programme on Bagasse Based Co-Generation

Based on the recommendations of the task force, the MNES had launched a National Programme on Bagasse Based Co-Generation in January, 1994.

The main objectives of the programme are:

I. Creation of conducive conditions for adoption of high efficiency co-generation schemes in Indian sugar industry leading to early exploitation of the estimated potential of around 3500 MW for surplus power generation.

II. Development of indigenous capability and capacity of design, manufacture and operation of modern high efficiency co-generation plants.

III. Generation and supply of 300 MW of power through bagasse based co-generation in Indian sugar industry during the VIII Plan Period.

The components of the programme include:

I. Non-recurring capital subsidy (upto 30% of the equipment cost or upto Rs. 70 lakhs/MW of net surplus power, whichever is lower) for projects that envisage the generation of at least 5 MW of surplus power using boilers which generate steam at a pressure of 60 bar or higher. However, the maximum amount of subsidy to any cogeneration project will be limited to Rs. 60 million in case of existing sugar mills. In case of new sugar mills the subsidy will be at half of the above mentioned levels. A maximum of 12 demonstration projects are proposed to be taken up in sugar mills in major sugar producing states in the country. Out of the 12 projects planned to be supported under the programme, 9 will be in the cooperative/public sector. The projects will be taken up in those states which have announced a general policy for purchase of surplus power generated.

II. Soft-semi commercial loans for projects which envisage production of at least 4 MW of surplus power. Under this activity MNES will provide grants (upto max. of 20 lakhs/ MW of surplus power) to FIs like IDBI, IFCI, ICICI, NCDC, IREDA, etc. to enable them to reduce interest rates by 2% on their lending (upto 60% of the project cost) to eligible sugar mills.

III. Assistance in the form of grant to the extent of upto 50% of the cost of preparation of Detailed Project Reports, subject to a maximum of Rs. 4.0 lakhs.

IV. Support to Research and Development projects in the areas of bagasse drying, bagasse storage and baling, fluidized bed combustion and gasification of bagasse.

V. Support to information dissemination, training and assessment studies.

Under the National Programme, the following sugar mills have been sanctioned in-principle a subsidy for surplus power co-generation (See Table 8.3).

IREDA's Role in Promoting Bagasse Based Co-Generation

The Indian Renewable Energy Development Agency has been entrusted with the responsibility to promote bagasse based cogeneration as a part of renewable energy development programme. In addition to this, IREDA also finances cogeneration projects based on other biomass fuels. The norms of financing under this programme are as follows:

TABLE 8.3: **Projects Sanctioned under the National Demonstration Programme**

Sl. No.	Name of the Sugar mill	Mill Capacity	Net Exportable Power	Subsidy Amount (Rs. in Lakhs)
1.	EID Parry (I) Limited, Tamil Nadu	5000 TCD	10.76 MW	450.25
2.	Kothari Sugars & Chemicals Ltd., Tamil Nadu	3500 TCD	9.90 MW	268.35
3.	Malaprabha Coop. Sugar Mill, Karnataka	3500 TCD	18.10 MW	600.00
4.	Thiru Arooran Sugars Limited, Tamil Nadu (Unit # 2)	2500 TCD	9.20 MW	343.00

IREDA's Norms of Financing

Minimum Promoters contribution	25%
Term Loan	Upto 75% of project cost
Interest rate (inclusive of interest tax)	16.5%
Total repayment period (inclusive of moratorium)	10 years
Moratorium (Maximum)	3 years

IREDA has so far financed the following bagasse based co-generation projects (See Table 8.4):

IREDA is presently being appraised by the Asian Development Bank for a US$ 150 million line of credit for various renewable energy technologies including bagasse based co-generation.

IREDA has also commissioned an All India Study on Bagasse Based Co-generation through the Maharashtra Industrial & Technical Consultancy Organization (MITCON), Pune. The key issues that will be dealt within the study include generation of project pipeline for IREDA by identification of sugar mills with potential for surplus power generation using bagasse and to identify and evaluate key issues of bagasse based co-generation projects for quick and successful implementation of such projects.

TABLE 8.4: **IREDA Financed Co-Generation Projects**

Year	Promoter	Project Cost	Total Loan	IREDA Loan	Remarks
1993-94	Narmada Sugars Limited, Khargone, M.P.	505.00	378.75	200.00	3 MW captive co-generation plant, Fully disbursed. Project has been commissioned in February 95.
1994-95	Rajshree Sugars & Chemicals Ltd. Madurai, Tamil Nadu	1300.00	850.00	850.00	12 MW co-generation plant. Operation only during crushing season. Export of 7.5 MW envisaged.
1995-96	Dharani Sugars & Chemicals Ltd.	3815.00	2861.00	500.00	15 MW Cogeneration plant. Operational round the year with lignite as fuel during off-season. Export power during season: 8.27 MW & in off-season: 13.47 MW
1995-96	Thiru Arooran Sugars Ltd. (Unit # 1)	4590.00	3440.00	2000.00	19 MW Cogeneration plant. Operational round the year with lignite as fuel during off-season. Export power during season: 9.2 MW & in off-season: 16.7 MW

Private Power Environment in India

The Government of India has targeted a capacity addition to the extent of 2000 MW through renewable energy including 300 MW through bagasse based co-generation. In order to achieve this target of power generation, the Government of India has announced several measures to promote independent power production. These measures include:

— Upto 100% foreign equity participation.

— Flexible power sales arrangements in which private power companies can act as licensees, supplying and distributing power to a given region or as generating agencies supplying power to the grid at large.
— 4:1 Debt-Equity ratio on investments.
— Single window clearance through a new investment promotion cell.
— ROI of 16% based on 68.5% PLF.
— 30 years initial investment period, with subsequent 20 years extensions.
— 5 year tax holiday, with partial holiday through tenth year of project.

Fiscal Incentives by Government of India

The other incentives offered for this sector are as under:

Item	Description
Income Tax	
01. Depreciation	100% depreciation in the first year can be claimed for the following co-generation equipment: 1. Fluidized Bed Boilers 2. Back pressure, pass-out, controlled extraction, extraction and condensing turbine for co-generation along with boilers 3. High efficiency boilers 4. Waste heat recovery equipment
02. Tax Holiday	First 5 years - Nil, Next five years - 70%.
Customs Duty	Duty leviable for NRSE power projects of less than 50 MW capacity (under Project Import Category) is 20% *ad valorem*, this covers machinery and equipment component parts required for generation of electric power.
Central Excise Duty	Exempted for renewable energy devices, including raw materials, components and assemblies
Central Sales Tax	-do-
General Sales Tax	Exemption is available in certain States

State Government Policies for Co-Generation

Table 8.5 gives a summary of the incentives offered by various state governments in India:

Current Co-Generation Activities in India

Co-generation activities in Indian sugar mills were restricted to the production of power required for captive use in the sugar mill and steam for sugar production process. Since the late eighties, there has been an increasing awareness of the advantages of surplus power generation from sugar factories, and some of them have taken steps to install medium pressure boilers (42 kg/cm^2) to generate surplus power for feeding to their ancillary units or to the state grid.

Two such sugar mills are the Daurala Sugar Works, Uttar Pradesh and Shree Datta Sehkari Shakkar Karkhana, Maharashtra have been exporting nearly 1.5 MW of surplus power to nearby industrial units. Two other sugar mills in the cooperative sector in Tamil Nadu (the M.R.K. Cooperative Sugar Mills, Sethiathope and Cheyyar Cooperative Sugar Mills, both of 2500 TCD capacity) have installed co-generation plants having a power generating capacity of 7.5 MW in their premises. These sugar mills have been designed to export nearly 4 MW of power to the SEB grid.

Multi-Lateral Investment Programmes

- The United States Agency for International Development (USAID) is presently operating a 19 million US $, five year Greenhouse Gas Pollution Prevention Project which considered a part of U.S. contribution to the Global Environment Facility. The project aims at reducing greenhouse gas emissions by catalyzing Indian development of efficient coal and alternative bagasse/biomass energy approaches for power sector expansion within India's economic development programme. The project is meant to promote efficient power technologies that have not been demonstrated in India. The project has two major parallel components, namely.

  — Efficient Coal Conversion (ECC)
  — Alternative Bagasse Co-Generation (ABC)

TABLE 8.5: **State Government Policies and Incentives for Bagasse Based Co-Generation**

Description	MNES Guideline	Tamil Nadu	Karnataka	Maharashtra	Uttar Pradesh	Madhya Pradesh	Punjab
Power Wheeling Charge	2 % of energy fed to the grid, irrespective of the distance from the generating station	2% of energy generated for 3rd party & 2% for sister concerns within 25 km radius	6% of the energy generated for 3rd party. (Likely 5%)	Nil	12.5% of the energy generated	2% of the energy generated	2% of the energy generated
Power Banking Period/Charges	One year/Nil	2% of energy generated	6% of the energy generated	—	One year	Not Specified	Not specified
Buyback rate by SEB	Rs. 2.25/kWh (min.) with 5% escalation without any restriction on time or quantum of electricity supplied	Rs. 2.25/kWh, with 5% escalation every year for 5 years	Rs. 2.25/kWh	Rs. 2.00/kWh for export upto 4 MW Rs. 2.25/kWh for export above 4 MW (Above rates for 6 years.)	Rs. 2.25/kWh or at a rate which ensures a RoE of 16% and 68.5% PLF, whichever is lower	Rs. 2.25/kWh	Rs. 1.50/kWh

(Contd.)

TABLE 8.5 *(Contd.)*

Description	MNES Guideline	Tamil Nadu	Karnataka	Maharashtra	Uttar Pradesh	Madhya Pradesh	Punjab
3rd Party Sale	Allow to HT consumers within the state at mutually agreed rate.	Allowed to HT consumers at mutually agreed rate	Allowed	Not Allowed	Allowed to HT consumers at mutually agreed rate within a radius of 5 km.	Allowed to HT consumers at mutually agreed rate	Allowed at mutually agreed rates.
Capital Subsidy	Subsidy given to industrial units in backward areas.	Nil.	Rs. 25 lacs/ MW (export)	For export > 4 MW, participation in cooperative mills by MSEB-upto 60% of equity	Nil.	Not specified	As extended to other industries
Other Concessions	- Exemption from electricity duty - Exemption from demand cut upto 30% of the installed capacity	1. Exemption from electricity generation tax	1. Exemption from electricity tax for 5 years for captive use. (Other modes - 4P per unit)	—	Nil	1. Exemption from electricity duty for 5 yrs. 2. Demand cut upto 30% of installed	1. Exemption from electricity duty. 2. Demand cut upto 30% of installed capacity 3. Exemptions from sales tax

(Contd.)

TABLE 8.5 *(Contd.)*

Description	MNES Guideline	Tamil Nadu	Karnataka	Maharashtra	Uttar Pradesh	Madhya Pradesh	Punjab
	- Sales tax benefits applicable to new industrial units. - Infrastructural facilities similar to industrial estates.					capacity 3. Exemptions from sales tax and other concessions applicable to new industry	and other concessions applicable to new industry
Promotional Agency		TNEB	KSEB	MSEB	NEDA	MPUVN	PSEB

This programme has a project investment component of US$ 12 million and a technical assistance component of US$ 7 million.

The ABC programme focuses on options for off-season power generation in sugar mills, so that sugar mills can supply firm power year round.

The highlights of the ABC Investment Component (US$ 9 million), meant to buy down the cost of the commercial line to be established by the IDBI are as under:

— Aims at demonstration and commercialization of bagasse based co-generation plants that rely on supplemental biomass fuels rather than fossil fuels. The biomass fuels recommended are cane field residues, stored bagasse which has been saved by the adoption of energy conservation measures, biomass such as rice husk, rice straw, groundnut shells etc. and wood from energy plantations.
— This component will work, closely with the sugar industry to provide the industry with information, technical assistance and training on all feasible options.
— The component will also cover the incremental costs and risks associated with the initial adoption of technologies and development of skills and knowledge needed for the introduction and use of supplemental biomass fuels by bagasse co-generation systems.
— The ABC projects will have a debt-equity ratio of 3:2.

The objective of the ABC TA Component (US$ 3 million), is to provide the pre-investment support needed to help stimulate private sector investment in biomass/bagasse based co-generation projects.

In addition to the above project the Asian Development Bank is also in the final stages of firming up a US$ 150 million aid for financing renewable energy projects including US $ 65 million for co-generation in India for on-lending through IREDA. The loan programme is expected to commence in April, 1996. As a part of developing a financing programme, IREDA signed a Memorandum of Understanding with the Bank to carry out studies in specific renewable energy sectors. The study has been carried out at a cost of US $ 371,000 including a contribution of US$ 354,000 from the Bank.

Concluding Remarks

Several studies carried out have established that significant potential exists in India for Bagasse based Co-Generation. Further, the interest shown by the Government of India, the State Governments and international funding agencies have provided the necessary thrust to encourage sugar mills to become more and more conscious for energy conservation and venture into the area of high efficiency co-generation leading to export of surplus power. As one of the largest processors of cane sugar, Indian sugar industry is on the threshold of making a visible penetration in power generation sector.

PART FOUR

CO-GENERATION

9

COGENERATION IN STIRLING ENGINE Gasifier Systems

—M. SUDARSAN AND R. BALASUBRAMANIYAN

Stirling engine-biomass gasifier systems are quite appropriate for remote rural area power generation or water pumping applications, where the availability of conventional sources of energy like petroleum products or grid electricity is limited and where the availability of non conventional sources of energy is assured. Also the same systems may be quite competitive in the agro based industries where the agro waste which is nothing but a form of biomass, is available in abundance. In such cases, however improvement in system efficiency may be quite essential. But the possibilities to improve the engine efficiency are appeared to be limited and complex and hence an effort is made to improve the efficiency of overall system by cogeneration.

STIRLING ENGINES

Features in General

Stirling engine is basically an external combustion engine where different kinds of heat sources can be used for supplying the heat to the working fluid to enable the engine to develop the mechanical power and is conceived by the Stirling brothers Robert Stirling and James Stirling in 1816. The theoretical Stirling cycle consists of four processes namely Isothermal compression, isochoric heat addition, isothermal expansion and isochoric heat rejection. The most unique aspect of the Stirling engine is regenerator to abstract the heat energy from the working fluid during one part of the cycle and reject the same to the working fluid during another part of the working cycle. The working fluids till now tried are air, hydrogen and helium. There are three types of engine configurations based on the arrangements of the pistons and cylinders and they are alpha, beta and gamma configurations[6].

Air standard efficiency of hypothetical Stirling cycle is higher than those of Otto and Diesel cycles and equal to that of Carnot cycle. But in practice, the thermodynamic efficiency of Stirling engines is very low, when it is compared to that of diesel or petrol engines. This is mainly due to thermodynamic irreversibility in attaining the perfect regeneration in the Stirling cycle.

ST5 Stirling Engine

The engine is of beta configuration and is developed by M/s Stirling Dynamics Madras with collaboration with Sun Power Corporation , USA. A cross sectional view of the ST5 Stirling engine is shown in Figure 9.1.

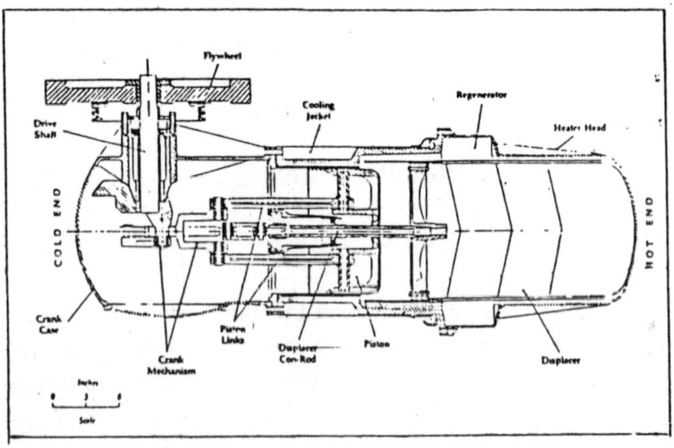

Fig. 9.1: Cross-sectional View of Stirling Engine

Design Specification[4,6]*:*

Rated Power Output	: 3.75 kW;	Working pressure	: 5 bar;
Rated Speed	: 720 RPM;	Working Fluid	: air;
Cylinder bore	: 30 cm	Stroke	: 10 cm;
Cooling water flow rate	: 15-25 l/min	Cooling water temp.	: 60°C max.
Drive Mechanism	: Crank Rocker	Configuration	: Beta

Engine Components[4,6]**:**

1. Heater	5. Cylinder	9. Fly wheel	13. Displacer
2. Regenerator	6. Crank case	10. Mechanism	14. Air drier
3. Cooler	7. Bearing case	11. Piston	15. Reed valve
4. Compressor	8. Crank shaft	12. Rings	

Novel Features[4,6]*:*

— ST5 Stirling engine is of beta configuration.

— The piston and the displacer are arranged in Tandem with phase difference of 85 degrees.

— Heater and cooler sections of the engine are provided with internal and external fins for better heat transfer.

— The regenerator material used in the engine is a commercially available material made up of knitted stainless steel wire.

— The swept volume in the expansion and compression are 6.70 and 5.80 litres approximately.

— A small builtin air compressor is incorporated to charge the engine during the idle run upto a maximum pressure of 5 bar.

DETAILS OF GASIFIER

The gasifier used is updraft gasifier designed and developed by CPRI. The gasifier is designed basically to suit for process heating applications and a flame guide is added to make the supply of heat uniform to the heater head of the Stirling engine. Assembled view of Stirling engine-gasifier system is given in Figure 9.2.

Specifications of gasifier[1]

Type	: Updraft & Multi fuel;	Output	: 100 kW (Max.);
Input	: 200 kW (Max.)	Firing Rate	: 30 kg/h (Max.)
Reactor Size	: 50 cm × 55 cm × 145 cm	Weight	: 75 kg.

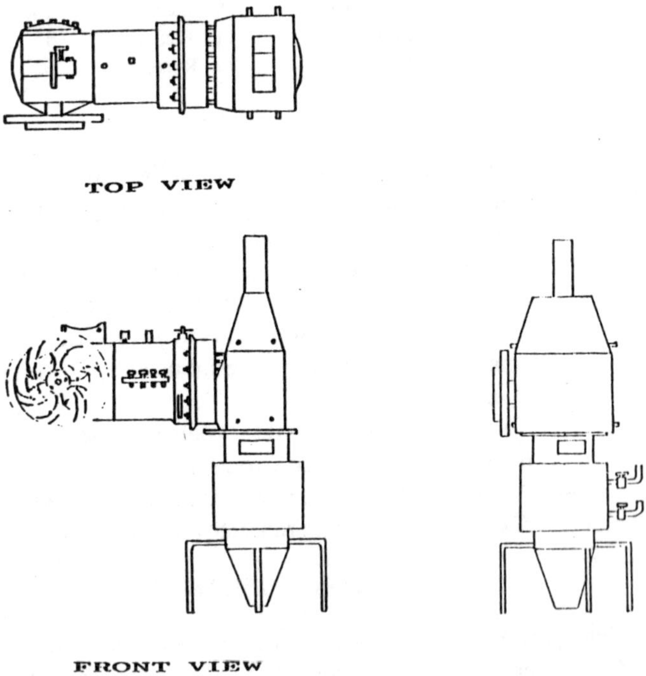

Fig. 9.2: Views of Assembled Stirling Engine-Gasifier System

Novel Features[1]

— Reactor of the gasifier is provided with water jacket of 20 litres capacity. This water will be converted into steam by utilising the heat lost to the walls of the reactor during gasification by radiation and convection. The superheated steam is injected into the reactor by six nozzles placed radially beneath the grate and aids the gasification process.

— It operates on natural draft and external power supply can be avoided.

— It has a high specific gasification rate of about 370 kg/h/sq.m.

— Fuel fluidization is also possible when fuel particles of smaller diameter are fired.

— The gasifier accepts wide range of biomass fuels.

COGENERATION

Since the maximum temperature of the working cycle is 650°C, the temperature of the flue gases that are coming out of the Stirling engine heater head should be greater than 650°C. Also it is clear from the Sankey energy balance diagram given in Figure 9.4 that the amount of heat that is lost to exhaust gas is around 25 kW. Hence the energy of this quantity at a temperature of 650°C is conducive to utilise for process heating applications such as tobacco curing, rice parboiling, pepper drying and copra drying etc. The variation in the overall system efficiency in general with the amount of the heat recovered from the exhaust gases of Stirling engine is given in Figure 9.3. Typical energy consumption and the process in brief are explained in the following paragraphs.

Parboiling of Rice

In small scale, the harvested paddy is initially soaked in the water for a period of about 8 hours. Then the soaked paddy is boiled in perforated tank in open pan and then it is sun dried and milled to separate husk and rice.

In large scale operations, the paddy is fed to the conical tanks where it is dipped fully in hot water at 80 to 85°C and continuously heated for 6 - 8 hours[2]. Then the paddy is transferred to perforated

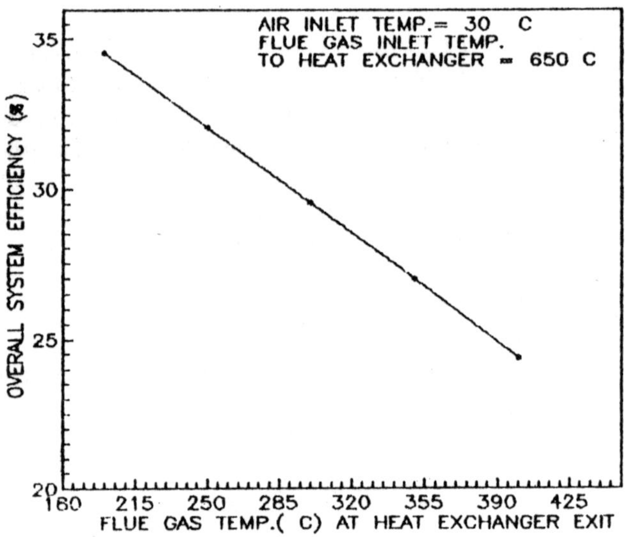

Fig. 9.3: Variation of Overall Efficiency with Flue Gas Temperature at Exit of Heat Exchanger

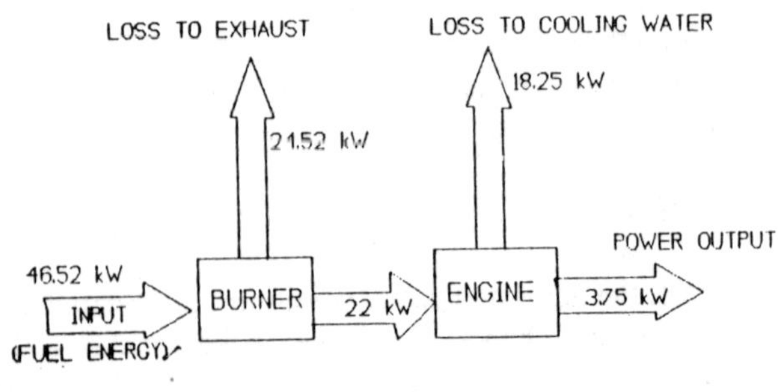

Fig. 9.4: Sankey Diagram for Stirling Engine-Gasifier System

trays for drying where the hot air at about 120°C is passed through the trays for about 5 hours. Afterwards the paddy is milled to obtain rice, broken rice, bran and husk.

The average energy consumption in parboiling of rice is 1758 kJ per kg. of raw paddy[2] processed and in other words for processing of a tonne of raw paddy the power required is about 48.8 kW. Hence we can say that about 250 kg of raw paddy can be processed with the energy in exhaust gases of Stirling engine, where it is assumed only 50% of the energy can be utilised. In case the energy in exhaust gases is solely utilised for supply of hot air the heat is sufficient to dry about 4 tons of raw paddy. But some modifications may be required in the size of the tanks and some piping layout. The advantage of the system is the power generated by the Stirling engine gasifier systems reduces the electrical power requirement of the unit.

Tobacco Curing

Tobacco curing is carried out in four phases: the yellowing stage, the colour fixing, leaf drying and stem drying. In a tobacco curing cycle, the relative humidity varied from 85 to 23 percent, the temperature is varied from 32 to 73°C[3]. The specific fuel consumption for total curing process varies 4 to 10 kg. of fire wood per kg of cured leaves. The average consumption of fire wood for total process is about 23 kg/h. The stagewise energy consumption is about 13 kg/h in yellowing stage, 32 kg/h in colour fixing stage, 54 kg/h in leaf drying stage and 16 kg/hr in stem drying stage.

Total time of drying varies from 90 to 110 hours[2] in which yellowing is for 35 to 40 hours; colour fixing is for 15 to 20 hours; leaf drying is for 15 to 20 hours and stem drying is for 25 to 30 hours. In energy terms, about 145 Megajoules are to be consumed to get one kilogram of cured leaves. The required capacity of the heat source will be about 115 kW for the drying unit capacity of 310 kg. of cured leaves. So the energy available in the exhaust gases of Stirling engine gasifier system can be used in systems of lower capacities of about 15 to 20 kg of cured leaves. Alternatively it can be used only for one of the four curing processes in relatively larger systems. However it also involves some modifications in drying system to connect to Stirling engine gasifier system.

CONCLUSIONS

— Stirling engine-gasifier system efficiency can be improved by cogeneration.

— Auxiliary power requirements of the drying plants may be reduced or replaced by incorporating the Stirling engine-gasifier system.

— Detailed process analysis is required before incorporating Stirling engine-gasifier system.

— Cost analysis pertaining to the necessary modifications is also essential.

NOTES

1. Development & Performance of a Novel High Rate Updraft Biomass Gasifier A Technical Report of CPRI.
2. J.S.Juneja, "Wood based energy system in rural industries and village applications, India," Field Document, Food and Agricultural Organisation of the United Nations Bangkok, July 1989.
3. Development of high efficiency gas-air heat exchangers to match updraft biomass gasifier, Technical Report No. 211, Central Power Res. Inst., Bangalore.
4. S. Seetharamu *et al.*, " Some experimental Studies on Performance of ST5 Stirling Engine with CPRI updraft Biomass Gasifier," *Journal of Renewable Energy*, Part III, pp 2096-2098.
5. M.Siddhartha Bhatt, "Utilisation Systems for Thermal Energy from Biomass Gasifiers," Proceedings of workshop on "Biomass Gasifier Technology- Recent trends in Process and Applications," Trivandrum, pp. 108-134
6. "Analysis and Experimental Studies on ST5 Stirling Engine" A technical report of CPRI, Bangalore.
7. M. Siddhartha Bhatt *et al.*, "High Efficiency Solar Thermal-Biomass Hybrid Agro-Industrial Drying Systems," Proceedings of National Solar Energy Conference, Nov. 26-27, 1994; pp. RA 5.1-5.5.

PART FIVE

BRIQUETTING

10

DECENTRALIZED IMPROVED CHARCOAL PRODUCTION FROM BIOMASS FOR INTEGRATED UTILIZATION

—R.C. MAHESHWARI

Biomass due to its inherent capacity to store solar energy and its amenability to subsequent conversion, is a renewable source of energy. However, its bulky nature, high moisture and tar content prohibits its direct application in a number of cases (FAO, 1983, 1985). In India about 80% of energy requirements are derived from biomass either directly or after conversion, as shown in Table 10.1, Fuel wood constitutes 68.5% of energy in the rural sector (Maheshwari & Prabhakar 1984).

TABLE 10.1: **Use of conventional and non-conventional energy in rural India**

Source	Consumption (%)
Non-conventional	
Fuel wood	68.5
Dung cake	8.3
Agro-waste	3-4
Conventional	
Oil	16.9
Coal	2-3
Others	3-4

Bioresource Technology 0960-8524/92/S05.00 © 1992 Elsevier Science Publishers Ltd., England. Printed in Great Britain.

Charcoal has been important to man for 10,000 years. Because of massive deforestation in some less developed countries, there is renewed interest in techniques of charcoal production which can produce higher yields or use unconventional biomass sources. 'Biocoal' is a name for biomass that has been treated thermally and mechanically to improve its properties for fuel use. It can be heated to intermediate temperatures to drive off water and make it hydrophobic, or made into charcoal. It can be pelletized to improve its use in combustion devices and reduce cooking costs.

Recent work in France has focused on making 'torrified wood' by heating biomass to 200-280°C (Antla, 1990). Commercial plants have been operated in India to produce 'Paru', by roasting biomass to about 300°C and then compacting it into briquettes for fuel purposes. A decentralized approach to biocoal production can fulfil area-specific requirements and reduce transportation costs. A number of surveys have shown that biocoal is being made through traditional methods with considerable waste of bioresources, e.g., in Rohtak and Gurgaon in Harayana, Saharanpur in Uttar Pradesh and Vishanpur in Bihar.

The processing of biomass to biocoal also reduces the moisture and tar content and produces clean fuel (Foley, 1986; White, 1983; Raintree, 1983). However, in order to check forest degradation, maintain an ecological balance and protect the environment, the technology requires some improvements. This will not only harness maximum energy from the bioresource but will also reduce pollution through byproduct utilization (Walker *et al.*, 1971; Crystopher, 1979; Vimal, 1984). In addition to this the technology is labour intensive and thus can reduce unemployment in rural areas to some extent.

The biomass is classified in 5 categories for charcoaling.

(a) Husk Type : Rice husk, saw dust etc.
(b) Leafy Type : Sarkanda leaves, paddy straw etc.
(c) Stem Type : This includes stems of agriculture crops like, Arhar, Cotton, Jute, Sarkanda and branches and Twigs of 1-3 cm thick.
(d) Woody Type : Logs of Eucalyptus, Acacia, Prosopis, Neem, Mango, Baniyan, Och tree etc., are suitable for fixed kilns and earth pit kiln.

(e) Shell Type : Materials like coconut shell are useful as raw materials in drum type and metal kilns.

CHARCOAL MAKING METHODS

The charcoal making methods have been classified into three types, viz earth or pit kiln, fixed kiln, and portable kiln, including drum type.

Earth or Pit Kiln

The earth or pit kiln method is very old and is widely used in many countries. Earth mound is generally preferred over pit, as the earth mound can be used over and over again, whereas pits tend to be used a few times and then new ones have to be dug to follow the timber resource. Also, where the water table is close to the surface or drainage is poor, pits are not practical, Repeated digging of pits also disrupts cultivation of crops and pasture. Hence, earth mounds are advantageous and are practical in agricultural zones. Earth mounds of different capacities can be constructed. For a mound of capacity 30 m^3, a base of at least 3 m radius is needed. Likewise, for mounds of capacities 90 m^3 basis of 4 m and 5 m radii are needed. Logs ranging from 20 to 40 cm diameter are used in these kilns. The whole cycle of production of 100 m^3 capacity is up to 24 days. The yield obtained in 100 m^3 capacity mounds is 120 kg/m^3. Yields of 30-40 kg/m^3 are very common. The main disadvantages of these kilns are longer period for production and difficulty in controlling air. Those who want to make occasional money (like farmers) can go in for small capacity mounds. These are not economical, for they need large logs of wood and give upredictable yields. These earth mounds can be used for industrial purposes and are useful in hilly regions. These are not useful for agricultural residues like paddy straw, rice husk etc.

Fixed Kilns

Fixed kilns are made out of brick, cement or clay. There are three important designs of this type.

Argentine Kiln

The shape of the kiln is hemispherical with a dia of about 6 m, the size of the bricks is 0.24 m × 0.12 m × 0.06 m. To construct a kiln, a total number of 5500-6000 bricks are required, making allowance for breakage. The kiln has two doors diametrically opposite each other. Fuelwood to be utilized is cut to 1.00-1.30 m length with a dia of 0.05. The kiln can be loaded with roughly 30 tonnes of air dry wood (25% moisture content). The kiln needs two men to operate and the operation schedule requires loading 6 (h), burning (6-7 days), purging (1-2 days), cooling (3-4 days), and unloading (3-4 days).

A total of 13-14 days should be adequate to complete a cycle to produce 9-10 tonnes of charcoal with a kiln of 7 m dia (7.5 t with a kiln of 6 m dia).

A battery of 10-14 kilns is needed for profitable operation. The disadvantages in these kilns are that they are expensive and agro-residues and other leafy biomass cannot be used as raw material for making charcoal. Only two charges can be obtained in a month.

Brazilian Beehive Kilns

These are circular, with a domeshaped roof and are built of ordinary bricks. The circular wall is totally in contact with the outside air. The Brazilian beehive kiln has the following dimensions; dia 5 m; nominal kiln volume, 48.94 m^3; effective kiln volume, 45.31 m^3; air inlets, 18; smoke stacks, 6; outlet ports, 6; emergency outlet ports, 50; and bricks required, 8500.

Two men can build a kiln in 8 days. The life span of the kiln is up to 6 years. Carbonisation time is 9 days, with a production of 5 tonnes per cycle. A battery of 7 kilns is needed to make the enterprise profitable; in addition, access roads, storage space for charcoal and wood, drainage facility, etc., have to be provided, these kilns are ideal in forest areas.

The Missouri Kilns

A typical Missouri kiln ranges in volume from about 4 to 350 m^3. The kiln is rectangular in plan, about 7 m wide and 11 m long. The walls are about 2.5 high. The maximum height of the vaulted roof is about 4 m above the floor. The walls are about 250 mm thick. The cycle time of Missouri kiln in warm climate is about a month made up as follows: Loading (3 days), burning (7 days), Cooling (21 days), and unloading (2 days), the total time is 33 days. The kilns are grouped in batteries of 3-6 or more. Its disadvantages are its high cost due to large quantities of steel and concrete used and its immobility. Unlike brick kilns, it cannot be demolished and rebuilt. Hence, a ten year wood supply must be available within the economic haul distance of any group of kilns. As the kiln is an engineering structure, it requires skilled labour and any laxity or lapse leads to complete collapse of the roof.

Portable Kilns

TPI design

The kiln consists of two interlocking cylindrical sections and a conical cover. The cover is provided with four equally spaced steam-release ports, which may be closed off with plugs as required. The kiln is supported on eight air inlet/outlet channels arranged radially around the base. During charring, four smoke stacks are fitted on to alternate air channels.

Drum type

This production method is based on 45 gallon oil drum designed to have air ducts, let into the kiln wall, which are sealed with plugs as the level of charcoal formed by the continuous fueling rises within the kiln.

The method is convenient for small-scale village production or for individuals wishing to utilise limited supplies of wood to provide charcoal for their own use. Indirect burning is recommended for agro-residues like rice husk, straw, etc.

OPERATION OF MUD BRICK BEEHIVE KILN

Carbonisation mainly depends on three factors:

1. Kiln design
2. Fuel wood grading
3. Kiln operation

The principles of carbonisation for mud and brick beehive kilns are identical while portable kiln has different firing system.

Apparently both are similar. Only method of construction techniques and materials are different but the design is same in both the cases.

Both kilns need hot air from burning firewood in the fire box chamber to heat up fuel wood until self carbonisation starts.

The carbonization of fuel wood will continue and the firing chamber will be reduced to limit the amount of air in the kiln. Observation of smoke helps to explain the progress of carbonization. The smoke will be first white thick with odour at the beginning and become thinner gray and then blue and finally transparent clear on complete carbonization.

PREPARATION OF KILN LOADING OR STACKING

For carbonization the kiln should be well prepared to the following manner:

(a) Clean the kiln floor from charcoal fines, tar, dust and soil from the floor and chimney side.

(b) Clean and repair all chimneys. Tar and carbon must be removed from the chimneys. The chimney wall may be repaired due to condensed water, acids, methanol etc., coming out on carbonization resulting the cracks.

Loading

For smooth, faster and efficient carbonization, the stacking of fuelwood in proper way is very essential. Fuelwood of equal size were piled vertically at the lower half of the kiln with thinner end to the floor. It will help draft circulation at the floor and reduce brand production. Bark may be removed from the fuelwood and carbonize separately in small kiln because combination of both may stop draft in some area of the big kiln also. A space of 5" × 5" is left at a smoke inlet of chimney or accelerating hole.

Large and green fuelwood are piled on the upper half of the kiln. It reduces time of operation and amount of brands. Large and green wood consume more fire wood to evaporate water from it. Carbonization always start from front and then top to bottom. Thin and dry wood at the bottom requires less direct wood burning.

Few holes (1" × 2") are kept at 12-18" distance on the kiln for spreading the hot gases from firing chamber uniformly. Larger holes in fuelwood piling are kept and more near at fire box chamber to allow hot gas. Fuelwood near the firing chamber is piled loosely in order to pass hot gas into the kiln easily.

Firing in the Firing Chamber

1. The fire wood is consumed at 2-3 Kg/hr. Hot air is sucked into the kiln to replace cold air which comes out of the kiln by accelerating hole. Kiln temperature is kept higher with respect to the temperature of vapour of smoke. Smoke from the hole gets thicker and hotter and contains vapour. Sometimes, smoke appears at chimnies.
2. Close the accelerating hole temporarily with brick till the kiln temperature reaches about 120°C. The smoke coming out from both chimneys thoroughly and smoothly for one hour before sealing the hole permanently with soft mud. It is reopened in order to get more heat in the kiln if smoke gets thinner earlier. Draft of the kiln is reversed after closing the hole. It conserves energy and heat moves downward to the kiln.
3. Fire wood 1-2 Kg/hr is consumed after closing the accelerating hole. The smoke is thicker white and smells of acid and methanol similar to hot spicy odour.
4. Reduce 2/3 of firing box after appearing white thick smoke and spicy odour for the 2 hours. It continues 15-30 hours depending of moisture content and size of fuelwood. The temperature is kept 68° to 75°C to this period kiln temperature remains 250° - 350°C.
5. If back pressure from flammable gases appears at firing chamber. It reveals that the carbonization is too strong and fast. Hence space of firing chamber is reduced until back pressure ceases. Sometimes cracks develop due to high temperature. They are sealed with soft mud.
6. Gray smoke turns to blue for 2-4 hours. During transition point, tar is observed on chimney surface. Tar sticks on the finger on touching. The tar is dried and hardened at the end of this period and smoke colour turns blue to clear. A first chimney is closed by a soft mud.
7. Blue smoke in other chimney continues till the tar is dried

and hardened on the smoke is clear. Space of firing box is closed before closing the last chimney and the kiln is completely sealed with loose soil, blocking all the air channels.

8. The kiln is cooled till room temperature. For speedy cooling it is rubbed with mud 2-3 times in a day.
9. The kiln is opened ensuring that the surface of the kiln is cold to touch. The charcoal thus obtained is then packed into bags, weighed and stored.

USE OF CHARCOAL

Wood is burn readily to produce heat, the heat value from charcoal is approximately twice as great as that from the same weight of wood. Moreover, charcoal burns with very little smoke and can be used in smaller or more efficient stove than those generally used for burning wood.

(a) *Household Cooking*: Charcoal is smoke free, noiseless, light in weight, convenient in use and provides the necessary heat for the preparation of meals as well as for hot water and social events. It is most widely used by food venders and restaurants.

(b) *Gas Generator*: It is used in producer gas unit for heating the houses, generating electricity and direct lifting the water through pumps.

(c) *Industrial Application*:

1. Water purification.
2. Purification of industrial wastes.
3. Fuel in foundry cupolas, electrodes.
4. Metallurgy for smelting and sintering iron ores.
5. Hardening of steel.
6. Production of Ferro-Silicon and pure Silicon.
7. Chemical industries for production of carbon disulphide and sodium cyanide.
8. Activated carbon and filter industry.
9. Lime and cement industry as fuel.

ECONOMICS OF CHARCOAL KILNS

	Description	Portable Metal kiln (2 kilns)		Brick Beehive Kiln (5 kilns)		Mud Beehive Kiln (5 Kilns)	
A.	Cost of Kilns two tonnes each (fixed type)	Rs. 12,000 each	24,000	Rs. 1500 each	7500	Rs. 650 each	3250
B.	Cost of Production per month	4 tonnes each time 15 runs		10 tonnes each time 6 runs		10 tonnes each time 6 runs	
(i)	Raw material cost Rs. 500/- per ton		Rs 30,000		Rs 30,000		Rs 30,000
(ii)	Labour at a rate of Rs. 20 per day (two persons)		Rs 1,200		Rs 1,200		Rs 1,200
(iii)	Packaging bags (Rs. 2/bag)	720 bags 28% yield	Rs 1,440	720 bags 30% yield	Rs 1,440	768 bags 32% yield	Rs 1,536
(iv)	Transport charge Rs. 800/- per truck		Rs. 1,600		Rs. 1,600		Rs. 1,600
(v)	Interest on fixed cost 16% per annum		Rs. 307		Rs. 100		Rs. 43
(vi)	Depreciation on fixed cost (20% per annum)		Rs. 383		Rs. 125		Rs. 54
(vii)	Interest on monthly investment (16% per annum)		Rs. 466		Rs. 460		Rs. 469
	Total Investment	Rs. 35,396		Rs. 34,925		Rs. 34,902	
C.	Charcoal Cost Rs. 2000/- per tonne		Rs. 36,000		Rs. 36,000		Rs. 38,400
D.	Net Profit (C—B)		Rs. 606		Rs. 1,075		Rs. 3,498

11

BIOMASS BRIQUETTING
An Indian Perspective

—S.K. MISHRA, P.V.R. IYER AND P.D. GROVER

India produces huge quantities of agro residues but they are used inefficiently causing extensive pollution to the environment. About 260 MTY of agro residues are generated in India out of which about 100 million tonnes remain unutilized[1]. The major residues are rice husk, coffee husk, coir pith, jute sticks, bagasse, groundnut shells, mustard stalks and cotton stalks. Sawdust is also available in certain parts of country. The direct burning of loose biomass in conventional grates is associated with very low thermal efficiency and widescale air pollution. The conversion efficiencies are as low as 40%. They also cause problems in handling, transportation and storage. Briquetting is one method which can take care of these pollution problems while using this important industrial/domestic energy resource. Normally bulk density of loose biomass is in the range of 0.05-0.2 g/cm^3 and can be densified to briquettes of density 1.1-1.4g/cm^3. These briquettes are easy to handle and eliminate problems of transportation and storage. The present way of transportation of loose biomass causing environmental hazard can be reduced if briquetting factories are encouraged in and around the major producers of agroresidues like rice mills, oil mills and agro-processing industries.

Biomass densification which is also known as briquetting of biomass has been practiced for many years in several countries. Historically biomass briquetting has been dependent on two main high pressure technologies, i.e. the reciprocating ram/piston press and the screw press. Both technologies have their merits and demerits but one fact is universally accepted that the screw pressed briquettes are far superior to ram pressed solid briquettes in terms of their storability and combustibility. The piston presses are of majority in use in India. Small screw presses of capacity 100 kg/hr are also in operation but they are very few in number. These presses are of no significance in India considering the huge amount of biomass residues. In the case of piston presses, numerous problems were encountered resulting in more instances of widescale failure than a few isolated success stories. But due to some promotional efforts by IREDA (Indian Renewable Energy Development Agency), the entrepreneurs still hope to use this technology to make some reasonable profits and pay-back their loan. Although the importance of biomass briquettes as substitute fuel for wood, coal and lignite is well recognized, but with numerous failures of briquetting machines in India, their extensive exploitation has been considerably retarded.

The trial of a screw press technology (Shimada : Europe) by Indian Institute of Technology, Delhi under a R & D project[2] in collaboration with University of Twente, the Netherlands has signalled an arrival of new technology in India. The success of this screw press technology under a commercial scale in utilizing various kinds of agro-residues along with sawdust has again enthused new entrepreneurs to invest in this sector. The two major impediments for the smooth working of the screw press identified were the high wear of screw and comparatively large specific power consumption. The mechanical wearing was enormous on the screw by using rice husk. The screw could last only 15-20 minutes in briquetting rice husk. The incorporation of well researched biomass feed preheating as a potential parameter for prolonging the standing screw life and reduction in specific power consumption of the machine was the focal activity in furthering improvement of this technology. In addition to selecting appropriate hardfacing alloys for screws, a thermic fluid based biomass preheater was engineered, installed and operated during this project. The results obtained were very favourable and much more than expected. The standing screw life increased drastically (40-44 hrs for sawdust, 31 hrs for rice husk), the power consumption decreased and the rate of production increased by 30 percent over the machine rated capacity. With this development the screw press technology has now become technically viable.

Briquetting Technology

Broadly, briquetting technology can be classified into two categories: low pressure and high pressure technologies. Low pressure briquetting technology usually employs a pressure less than 50 bar with or without using a binder. These low pressure technologies can handle raw material with moisture content as high as 30%, sometimes more than that. This method is applied for briquetting green plant wastes or highly moist residues such as coir pith. These briquettes need drying before they can be used as fuel. In all these cases actually briquette is 'formed' rather than 'densified. The density of briquettes vary between 0.5 and 0.7 g/cm^3 which is achieved after drying the briquettes to an acceptable quality. Still they have little machanical strength to withstand rough handling during combustion. Hydraulic press is generally used producing these kind of briquettes. But the birquettes are of no use as far as its commercial application is concerned. The same technology can be used to press raw material with the addition of a binder which adds extra cost to the product. By using a binder, one has to be careful with the moisture content of raw material. Otherwise the while purpose of briquetting is lost because of inferior quality of briquettes. But the binder can be used successfully in a high pressure technology, which employs pressure more than 500 bar. This is also possible to obtain briquettes of good quality by not using a binder in a high pressure technology. Piston press and screw press are two high pressure briquetting technologies which are of commercial importance.

In a country like India where huge amount of agro-residues is available, a technology capable of handling wide range of biomass and which can give good quality briquettes without a binder, of reasonable cost to be competitive with wood, is present day's need. At present, piston presses are the only commercially available machines in the country. They are of 500-700 kg/hr capacity. In this technology the material is gravity or force fed into the compression zone where a moving piston forces the material through a slightly tapered die. The piston is driven by a large fly wheel with an electric motor. The taper angle and the length of the dia determine the back pressure needed to form the briquette. The shape of the briquette is invariably cylindrical with a real density of 1.1-1.4 g/cm^3. Though the briquette is hard and compact, the weakness lies in the segments (due to strokes) of the briquettes. They crumble during combustion and a reasonable quantity is lost under the grate. These briquettes are also not suitable for use in open fire places (initially produce lot of smoke due to incomplete surface combustion) and cannot be carbonized. This technology is known to be able to operate on a wide range of raw materials which is a positive point of course with the exception of rice husk which is abundantly available in India. But the maintenance of different heavy parts like crank and shaft, fly wheel, ram and die does not suit a new entrepreneur who is not very familiar with the technology. Wear of the die and piston is a frequent problem and they need more frequent replacement when more abrasive materials such as rice husk is used. Oil contamination with the raw material dust is another problem to comply with. In India many manufacturers produce this type of densification equipment which appears to have originated in Switzerland in the 30's and was originally patented by Haussman. Solar Sciences Consultancy Pvt. Ltd., is a leading private company in the manufacturing and installation of piston presses.

The screw press technology originally from Japan, uses a revolving screw (600 rpm) to compress the biomass against a heated die (280 °C) to achieve a briquette density of 1.1-1.2 g/cm^3. This briquette is highly of homogeneous nature with a central hole. The continuity in structure and the central hole both prevent it from crumbling during combustion. The central hole accommodates the volatiles and help the air for mixing for its complete combustion. This technology uses an octagonal die to produce briquettes of the same shape, diameters of 55-65 mm. Machines are available for import of 400 to 700 kg/hr which is of course different for different biomass. Another salient feature of screw press technology is its product which has a partially carbonized surface so obtained because of the heated die. This carbonized surface acts as a resistant to moisture and helps against absorption of moisture in ambient conditions. This is a very important aspect from storage point of view. The screw press can successfully densify rice husk after its grinding. Rice husk does not need drying which saves an extra cost on fuel.

In India screw press technology has not come into the fore. These machines are available only in Japan and Scotland, the brand name being Shimada (Japan) and Shimada (Europe)[3] respectively. In the past[4], a screw press technology imported from Japan failed miserably due to its lack of application in utilizing Indian biomass which otherwise a great success in Japan and Europe. However, research and development project by IIT Delhi in collaboration with University to Twente, the Netherlands has not only thrown light on the failure of the technology but also has improved it to suit Indian conditions. This project supplied a unique concept of preheating (on larger scale) which used a thermic fluid preheater to heat the biomass prior to its briquetting. In this method, biomass was preheated to soften it and enhance cohesion to further strengthen the binding mechanism. This resulted in increase in output, reduction in power consumption and increase in the life of contact parts like screw and die. Table 11.1 gives the results of different biomass on screw life and the corresponding tonnage of briquettes produced.

TABLE 11.1: **Data on briquetting of different preheated biomass**

Biomass	Screw life (hrs)	*Production rate (kg/hr)	Tonnage produced	Need to repair
Sawdust	44	360	15.84	yes
Rice husk (ground)	31	500	15.50	yes
Mustard stalks (ground)	12 (ic)	360	4.32	no
Groundnut shells (ground)	12 (ic)	480	5.76	no
Coffee husks (ground)	27 (ic)	600	16.20	no

ic : test is incomplete
* rated production capacity of the machine = 400 kg/hr

Economic analysis carried out in this project indicated that the screw repair cost was no more the key factor which was an economic hindrance for commercial application of this technology. The increase in screw life brought down the screw repair cost to as low as Rs. 35 per tonne of briquette produced. This is definitely a step forward in the briquetting technology which is facing technical as well as entrepreneurial sufferings since many years.

Promotional Aspects and Future Scope

The Indian experience in the failure of briquetting technology has nothing to blame in regards to financing or any other way of promotion. The more encouraging fact is that both MNES (Ministry of Non-conventional Energy Sources) and IREDA have shown grèater interest to improve this sector. Especially the efforts made by IREDA is commendable. They have instituted a technical back-up cell (TBC) to monitor and improvise the briquetting units in many ways. The proposal made by IREDA to achieve production of one million tonnes of briquettes in the next five years has again revived the interest among entrepreneurs, present and future. MNES is also trying its best to restructure the whole briquetting sector to give it a new direction. The constant meetings and interaction of IREDA officials with entrepreneurs have helped in revitalising the efforts. No doubt, the defaulted entrepreneurs have some hope now to compensate their past losses and can think of making their projects successful. At the same time people now have realised the need of a modern technology easy to operate and handle.

The screw press technology (Shimada : Europe) already having credentials in Europe is the best suited technology which has been tested in Indian conditions by the IIT project, as mentioned earlier. This technology is originally from Japan and has been successful in Japan and Europe on large commercial scale. Though dissemination efforts have been made in south-east Asian countries like Thailand, Malaysia, Indonesia, Taiwan etc., it is yet to emerge on large commercial basis. The reason being lack of proper infrastructure and technical efforts in this line. However, with the vast technical manpower in India, the screw press technology can make a headway. A method should be worked out to import this technology first and then subsequently manufacture the same in India under a joint venture. This is high time now to realise technological attributes, which play an important role in present and future cost and quality of the product. This has become a dominant criterion in product saleability. Successful integration of technology and business planning can utilize technology-based resources to create an overall business edge. Key technologies require at least 8-10 years time frame to develop which is now over for piston press. However, the use of piston presses should not be discouraged and technical support is essential for piston presses because of their large present use in India. On the other hand, if promoted the screw press technology will be able to build a base in India on prior technology and its present success through R & D efforts. This technology was successfully tried using sawdust, rice husk, mustard stalks, groundnut shells and coffee husk. With suitable mix, materials like bagasse pith, coir pith and tea waste can also be used for briquetting. The prevailing situation in at least for the last ten years has warranted a new promotional policy and investment procedure to give a launch and then a boost to this technology. This might start with some demonstration plants with the support from IREDA and Government of India. International cooperation is also possible in this line.

Market Potential

There is no dearth of market for briquettes. These can efficiently compete with other fossil fuels like coal and wood. Continuity in supply and confidence in the product are the two major considerations for marketing briquettes in India. In some cases the users of solid fuels are reluctant to switch over to this product because of their lack of exposure in utilizing briquettes and also are apprehensive of temporary shutdowns of their processes due to lack of expertise in handling briquettes. But those who have frequented using briquettes have found them excellent.

Small scale industries are the major marketing base for biomass briquettes. Rubber industries, textile dying units, leather processing units, small boiler units, tobacco processing units, brick kilns and the domestic sector use huge amount of wood and loose biomass for heating, drying of steam generation. Utilization of loose biomass of course should be totally discarded because of their inefficient utilization (less thermal efficiency) and producing dust and smoke pollution. Reduction on woodfuel utilization requires no further mention which has been now a global issue. Therefore, the whole idea is the use of loose biomass more efficiently.

Foremost is to identify small scale units throughout the country and educating the users about the benefits of briquettes. Some of the units already using briquettes can also help in this regard. The bottlenecks if any in burning the briquettes can be removed or improved by a technical group formed under any technical institute. This supporting group can mobilise and provide technical help for efficient use of biomass briquettes. This will act like a guarantee system for the manufacturers of briquettes who will try to increase their production to meet market demand and ultimately the use of huge amount of agro-residues and other biomass residues will be met. In this way, the market will gradually switch over to biomass briquettes. This will not only encourage manufacturers of briquettes but also the users who can ease their operation to their benefits. Many directions like manufacturing of machines, production and use of briquettes, reduction in consumption of wood and loose biomass, reduction in environment pollution and employment generation are possible from a single sector. So far, the prevailing situation in India can suggest that best co-ordination among users and manufacturers of machines as well as briquettes with a strong marketing base, along with a suitable government policy and direction will lead to a successful briquetting industry in India.

Conclusions

(1) A promotional base should be developed to launch screw press technology.
(2) Efforts should be made to educate users of briquettes in a way developing a market base for briquettes.
(3) Briquetting promises development in many ways for a country. Presently it needs support from Government of India from financial and promotional point of views.

NOTES

1. Annual Report, 1993-94, Ministry of non-conventional energy sources, Government of India.
2. P.D. Grover, Biomass briquetting : technical and feasibility analysis under biomass densification research project (Phase II), International conference in biomass briquetting, April 3-6, 1994.
3. H. Brandt. Densitech Pvt Ltd, Koninksweg 1-3, 7597 LW Saasveld, The Netherlands.
4. M.K. Srinivasan, MAV Industries, 320-1-A-1, Thiru-Vi-Ka-Road, Karur-639001, Tamilnadu.

PART SIX

GASIFICATION

12

PERFORMANCE STUDY OF JETROPHA OIL IN THE PLACE OF DIESEL IN BIOMASS GASIFIERS

—B. RAJA, V. NARASIMHAN, M. MURUGAN,
G. THIRUMURUGAN AND T.M. HARIDASAN

During the last ten years biomass gasifiers developed in the country have been put to various applications such as thermal, mechanical and electrical end uses in various industries[1]. In the traditional uses making the mechanical or electrical modes of applications, the gasifier is normally attached to diesel engine of appropriate capacity along with an alternator of suitable rating. Before allowing the engine to start running, diesel is applied as the starting fuel and steadily the diesel quantity is reduced by letting in the producer gas such that the system runs on dual fuel mode. Typically a diesel replacement of 60 to 80% can be achieved in the various gasifiers developed in the country. Simultaneously there have been also efforts in our country and abroad for finding out substitutes for diesel both for stationary and mobile applications. One such substitute happens to be the Jetropha oil which is available in the country in large quantities thanks to its proliferating and cattle repelling properties making it thus an ideal plant for developing the waste lands. Different investigators have measured its fuel properties in comparison with diesel and found to be a favourable substitute for diesel inspite of its higher viscosity and flash point.

In the present study an attempt is made to use Jetropha oil as the starting fuel in a diesel engine alternator system. The system then is allowed to run on dual fuel mode with Jetropha oil and the producer gas as the dual fuels. Comparison is made of its performance with a similar set-up using diesel and producer gas as the dual fuels.

Fuel Characteristics of Jetropha Oil

A non-governmental organisation named "SEVA" of Madurai was kind enough to give us the oil for the present test. They are promoting the production of this oil for soap manufacturing. Having seen its potential as a diesel substitute, they were eager to know its fuel characteristics. Eventhough the fuel characteristics of this oil are reported in literature[2], we have carried out such investigations on the specimen given to us for continuity and consistency. The relevant characteristics are given below : Table 12.1.

TABLE 12.1: **Physical properties of fuels**

Property	Jetropha oil	Diesel
Flash point	290 °C (* 246)	50 °C
Kinematic Viscosity	0.431 s	0.0251 s
Calorific value	9500 Kcal/kg	10170 Kcal/kg
Cetane number	51	46-50

Testing in the Engine system

The diesel engine used is a 5 HP, 1500 rpm Field Marshal Engine with the two modifications on:

(a) Fuel feed system such as filter, fuel injection pump and fuel injector,

(b) Engine injection timing.

The engine was run with diesel, and then diesel blended with Jetropha oil and finally with 100% jetropha oil. This is done to study the effect of the Jetropha oil on engine so as to ensure that the engine is no way affected before we start the experiment on dual fuel mode. The results are summarised in Table 12.2.

There is no drastic variations in exhaust temperature, RPM or in the duration for consumption of fuel eventhough the time span goes smaller as the percentage of Jetropha oil increases.

TABLE 12.2: **Engine performance using different blends of Jetropha oil**

Percentage of Blend	Exhaust temp. °C	Speed in rpm	Time for 50 cc consumption in minutes
100% Diesel	130	1509	7.38
10% J. oil + 90% Diesel	128	1508	7.36
20% J. oil + 80% Diesel	125	1509	7.31
30% J. oil + 70% Diesel	125	1523	7.21
40% J. oil + 60% Diesel	124	1525	7.11
50% J. oil + 50% Diesel	123	1534	7.19
60% J. oil + 40% Diesel	124	1536	6.59
70% J. oil + 30% Diesel	123	1544	6.53
80% J. oil + 20% Diesel	126	1543	6.25
90% J. oil + 10% Diesel	127	1545	6.13
100% pure Jetropha oil	129	1545	5.58

Performance study in a gasifier on dual fuel mode

The gasifier used is a 3.75 KW capacity of AEW make. To standardise its performance the dual fuel operation was done first with diesel and producer gas for various loads. The results are shown in Table 12.3.

Then the experiment was repeated with Jetropha oil replacing diesel for various loads with results in Table 12.4. The maximum diesel replacements was at 50% load (75.5%). The speed in RPM does not change drastically in either cases for various loads. The pressure drops at gasifier outlet and after cooler are considerably

TABLE 12.3: **Performance of Gasifier System with Diesel Producer Gas Dual Fuel**

Load in %	KW	Volt (V)	Amps (A)	Time for 50cc Fuel (sec.)	Time for 50cc Dual (sec.)	Repl. (%)	Speed rpm	Exhaust temp. °C	Pressure P1 (mm)	Pressure P2 (mm)	Temperature T1 °C	Temperature T2 °C
0	0	250	0	406	679	40.2	1550	200	14	13	223	48
10	.375	250	.8	358	612	41.5	1541	222	14	15	206	42
20	.740	250	2.5	308	738	58.2	1535	243	11	21	196	40
30	1.115	250	4.2	267	841	68.2	1521	293	18	22	191	39
40	1.475	250	5.7	230	721	68.1	1513	329	18	14	202	38
50	1.830	250	7.2	220	902	75.5	1515	355	19	23	195	37
60	2.225	250	8.5	184	702	73.7	1508	427	23	25	213	38
70	2.590	250	9.8	169	526	67.8	1500	416	19	28	199	38
80	2.960	250	11.0	154	308	50.0	1496	435	19	28	185	39

Fuel—diesel; P1—Gasifier outlet pressure; P2—Cooler outlet pressure

T1—Gasifier outlet temperature; T2—Cooler outlet temperature

TABLE 12.4: **Performance of gasifier system with Jetropha oil—Producer gas dual fuel**

Load in %	KW	Volt (V)	Amps (A)	Time for 50cc Fuel (sec.)	Time for 50cc Dual (sec.)	Repl. (%)	Speed rpm	Exhaust temp. ℃	Pressure P1 (mm)	Pressure P2 (mm)	Temperature T1 ℃	Temperature T2 ℃
10	.375	250	.8	239	318	24.8	1525	266	5	15	224	41
20	.740	250	2.5	235.5	412	42.9	1524	292	8	12	222	39
30	1.115	250	5.3	221	369	40.1	1569	309	8	16	215	38
40	1.475	250	6.5	199	453	56.0	1552	319	8	14	200	39
50	1.830	250	8.0	185	345	46.3	1533	334	7	15	198	38
60	2.225	250	9.2	173	310	44.1	1540	357	8	15	196	38
70	2.590	250	10.5	153	265	42.2	1535	397	7	15	188	38
80	2.960	250	12.0	139	169	17.5	1520	435	7	15	188	38

Fuel—Jetropha oil; P1—Gasifier outlet pressure;
P2—Cooler outlet pressure; T1—Gasifier outlet temperature;
T2—Cooler outlet temperature.

smaller when Jetropha oil is used in the place of diesel. The gasifier outlet temperature also does not show much variation when diesel is replaced by the oil. However on the cooler side the temperature is fairly steady for various loads when the Jetropha oil is used.

Conclusions

The preliminary studies as mentioned above indicates that Jetropha oil can be used in biomass gasifiers in the place of diesel as the starting fuel. The performance characteristics reveal that eventhough Jetropha fuel replacement by producer gas is less the overall nature of the engine performance is fairly satisfactory. The economics of using Jetropha oil, though not worked explicitly here may not be attractive now since its price is of the order of Rs. 14 per litre.

However, due to excessive cultivation and generation of oil this price may come down. More experiments are necessary by reducing the viscosity of the oil by suitable blending to ascertain the suitability of this oil in gasifier-engine systems.

Acknowledgements

The authors are thankful to Ministry of Non-conventional Energy Sources for the funding to Action Research Centre on Biomass Gasification where this work is carried out. They are also thankful to "SEVA", the NGO based at Madurai for supplying the oil in desired quantity for this sutdy.

NOTES

1. Annual Report, Ministry of Non-conventional Energy Sources 1994-95.
2. Jetropha Curcas. The Wonder Plant by Vinayak Patil and Kanwarjit Singh. Represented by Food and Agriculture Dept., Government of Andhra Pradesh, Hyderabad.

13

BIOMASS FUELED STIRLING ENGINE—SOME OPERATIONAL CONSTRAINTS

—N.S. RATHORE AND PRATAP SINGH

Stirling Engine is an appropriate prime-mover which can be used for different on-farm applications. It is suitable for outdoor installation with a considerably long life under adverse farm conditions. In addition, such an engine can be run on any type of combustible farm residue or waste, such as wood, rice husk, saw dust, wheat husk, etc., that can burn satisfactorily, giving it a multifuel capability and capacity to use renewable energy sources that could be generated at user location itself. Thus, it has the potential to be a good energy source for farming operations.

The currently available biomass fueled stirling engine with specifications as detailed in Table 13.1, is extensively used for on-farm applications in various parts of our country such as water pumping, chaff cutting and electricity generations.

The stirling engine was operated under both no-load and on-load conditions for water pumping, electricity generation and chaff cutting with a view to judge its operational convenience. During these trials, many limitations in the operational performance of the system have been observed, which are presented in this Chapter.

Operational Constraints

During operation of stirling engine for on-farm applications, few major problems were encountered. Table 13.2 gives some problems and their remedies for better improvements.

TABLE 13.1: **Specifications of the Available Biomass Fueled Stirling Engine**

S.No.	Specification		Particulars
1.	Type	:	Single acting, Single phase non-resonant, rigid beta coupling type biomass/fuelwood stationary hot air engine
2.	Fuel used	:	Rice husk, Saw dust & various agricultural residues in particulate form
3.	Operating temperature	:	650°C (maximum)
4.	Operating pressure	:	5 bar (maximum)
5.	Power output	:	4 KW at 650 rpm available at fly-wheel.
6.	Specific fuel consumption	:	2.5 kg/KW/h of rice husk.

TABLE 13.2: **Identified Problems and their Remedies in present Stirling Engine (Model : HAMARA ST-5)**

S. No.	Installation	Caused by	Remedy
1.	Engine Shape	- Presently engine is for stationary works. If require a mobile chassis for mounting over that, so that engine can be transported to field for different on-farm applications.	
	- Mobile chassis may yield excessive vibrations and develops cracks.	- Mobile arrangement	- Through suitable jacks. - Through vibration absorber. - Balanced chassis should be provided. - Material of const-

(Contd.)

TABLE 13.2 *(Contd.)*

S. No.	Installation	Caused by	Remedy
			ruction should be such that it can resist sudden shocks & excessive vibration.
2.	Feed Supply System	- It consists of a feed hopper fitted with an angle. It gets drive from the engine itself.	
	- Auger gets choked frequently	Presence of foreign matter or agglomeration due to moisture content of the feed.	- Pre-conditioning of raw feed to attend a particle texture suited to the existing system. - Provision of a screen for arresting objectionable matter. - Replacement through an alternative fuel supply system based on flutted roller mechanism/belt conveyor type.
	- Presently available feed supply system is not suitable for leafy biomass, wood chips, crop residue etc.	Design is suitable for material of particular form only.	- Alternative screw conveyor should be designed to accommodate wide variety of fuels.
	- Feed auger fails to supply required quantity of feed at maximum engine speed i.e. 500 rpm	Faulty auger mechanism and partly because of agglomeration of raw material.	- Auger should be adjusted for proper supply of raw material. - Supply perfect raw material only.

(Contd.)

TABLE 13.2 *(Contd.)*

S. No.	Installation	Caused by	Remedy
			- Supply higher calorific value fuel at maximum engine speed.
	- Drop in engine speed due to sudden increase in load & thereby decreasing fuel supply rate.	Poor response of the engine to fluctuating loads.	- Rating of fuel supply system be kept at the higher level required at the lowest level. - Flap valve mechanism be inserted to adjust the fuel supply.
3.	Combustion System	- It is provided with refractory bricks, a insulator (glass wool) is also provided over the combustion housing.	
	- Fire bricks line can develop cracks & start coming off.	Because of continuous combustion process long hours duration.	- Presently available refractory bricks have only 200 hours of operational age. - Scrubbing of combustion surface for ash removal also cause its damage.
	- Leakage of fuel gas from the joint between the heater head & the combuster.	Due to fuel with relatively high moisture content.	- Use dry fuel. - Place an asbestos rope packing at the joint.
	- Generation of heavy smoke.	Because of high moisture content in the feed material.	- Use dry fuel.
	- Lot of muscle power	Due to heavy-	- Combustor should

(Contd.)

TABLE 13.2 *(Contd.)*

S. No.	Installation	Caused by	Remedy
	required for moving the combustor during starting & closing of each run.	ness of combustor.	be redesigned with a lighter refractory material & adequate insulation.
	- Combustion with normally available leafy rice husk is not found clean.	Due to low combustion.	- Provision should be made for intermittent charging of wood for proper combustion.
4.	Cooling System	- Cooling is done by circulating water in the jacket, formed over the cold end of the engine by wrapping their gauged stainless sheet.	
	- Existing cooling system is not suitable for cleaning & proper operation.	Teething problem in existing cooling system.	- Replace through jacket, cooling system similar to automotive engine or self-contained cooling system.
5.	Starting Mechanism	- The Stirling engine is started manually by rotating the fly-wheel, after the heater head is sufficiently hot.	
	- Lot of muscle power is needed for starting the engine.	Inefficient starting mechanism.	- Design an alternate efficient starting handle.
		No means to ascertain the temperature of heater & cooler head.	- Design an adopter for accommodating the handle on fly-wheel shaft.
			- Install temperature measurement units at heater & cooler end.

Conclusions

The biomass fueled stirling engine could be useful for the purpose of pumping of water, post harvest operations, electricity generation and farm level demonstration in remote areas if its operation could be perfected in terms of reliability, efficiency and convenience. It has been observed that engine requires some fundamental changes in its present shape in feed supply system, combustion system, cooling system and in starting system. Further, there is present need to mount the engine on wheel, so that power could be transported to field for different on-farm applications.

REFERENCES

1. Proceeding of Bio-Energy Society V Convention & Symposium. Edited by R.N. Sharma, *et al*. (Bio-Energy Society of India, New Delhi) 1988.
2. Ross, A. "Stirling cycle Engine", Solar Engines, 1977.

14

SYNTHESIS OF PURE AMORPHOUS SILICA AND COMBUSTIBLE GAS FROM RICE-STRAW AND HUSK

—A. CHAKRAVERTY

Rice straw and rice husk contain about 15 and 20 per cent silica respectively. The biogenic silica present in these biomass is inherently amorphous (Chakraverty, A., 1989). The carbon free silica obtained from rice-straw and husk has opened a new era in its utilisation concept as it is an attractive and cheap source for manufacturing silicon of reasonably pure quality (Mishra, P.N., Chakraverty, A., and Banerjee, H.D., 1985). The present project has been undertaken with the following major objectives : (i) to study the thermal decomposition of rice-husk and straw by thermogravimetric analysis (TGA) and (ii) to study the effects of acid leaching of rice-straw and husk, furnace temperature, straw and husk-bed depth on the production of combustible gas and pure amorphous white silica.

MATERIALS AND METHODS

The straw of 1 cm in length and husk were thoroughly cleaned separately with distilled water and then sun dried. The dried materials were leached with 1 (N) HCl at 75°C for 1 h and washed with distilled water for several times to remove the HCl and then these were dried in the sun to bring down moisture content (m.c.) to 6-7% (wb).

Moisture contents of samples were determined by an air oven drying method (100 ± 2°C for 24 h).

The TGA of husk and straw were carried out in still air using a thermogravimetric analyser at a constant rate of heating, 10°C/min.

A vertical electric furnace equipped with thermocouples in conjunction with a temperature controller-cum-indicator, a temperature recorder and a digital multimeter and an energy meter was used for all experiments. It was specially designed to carry out thermal decomposition for the production of silica and gas. The furnace temperature was raised to the desired constant temperature and then either straw or husk was fed to the furnace to expose it to the stable furnace temperature with a natural draught. Combustion continued with natural draught till the entire mass was converted into white ash.

The structure of silica in the above white ash was examined by X-ray diffraction technique with the help of a X-ray diffractometer at a chart speed of 2°/min using CuK_a radiation.

RESULTS AND DISCUSSION

Thermogravimetric Analysis

The overall nature of the TGA of the raw and acid treated straw and husk shows that the mass loss occurred in three major stages namely, (i) drying; (ii) loss of volatiles and (iii) oxidation of fixed carbon (Figs 14.1 to 14.4). During 1st, 2nd and 3rd stages about 7-10, 64-68 and 13-19 per cent mass loss took place.

Combustible gas

It was observed that as the bed thickness increased, the flame temperature and the duration of flame also increased. It was also found that as the furnace temperature increased, the flame temperature also increased for both straw and husk. The flame temperatures of the gas during their gasification were about 650° and 750°C corresponding to the furnace temperature of 450° and 500°C at a bed depth of 20 cm.

Pure amorphous silica

At a furnace temperature of 500°C the rice-straw and husk did not become milky white even after long exposure. It was necessary to pertreat them with 1 (N) HCl at 75°C for 1 h combustion to remove their metallic impurities for the production of pure white ash within a

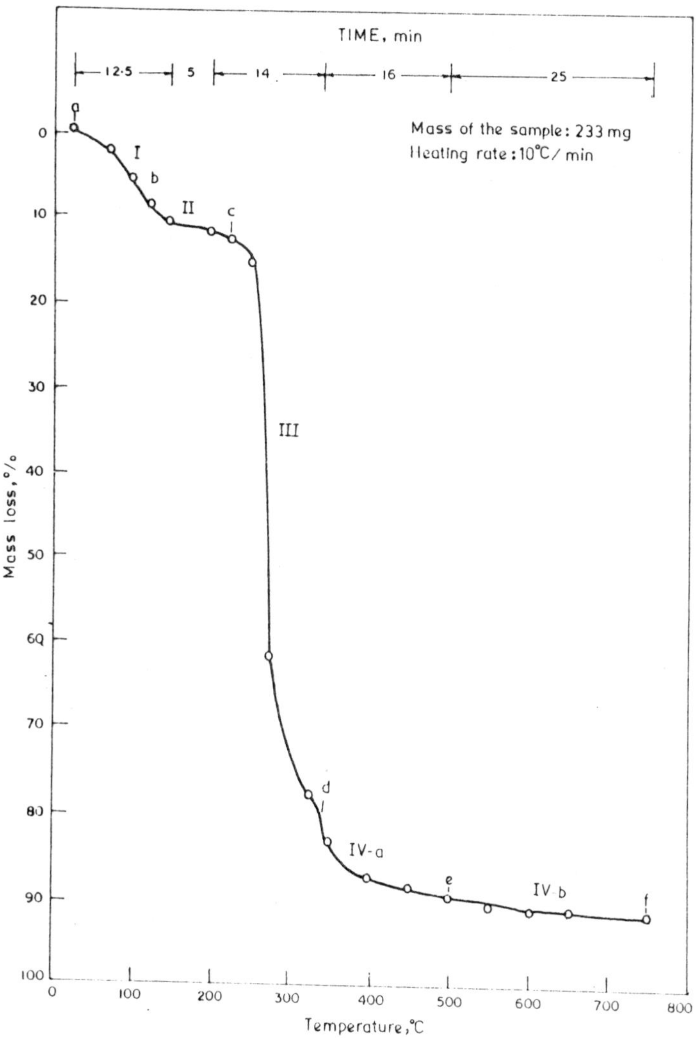

Fig. 14.1: TGA Curve of Ground Raw Straw

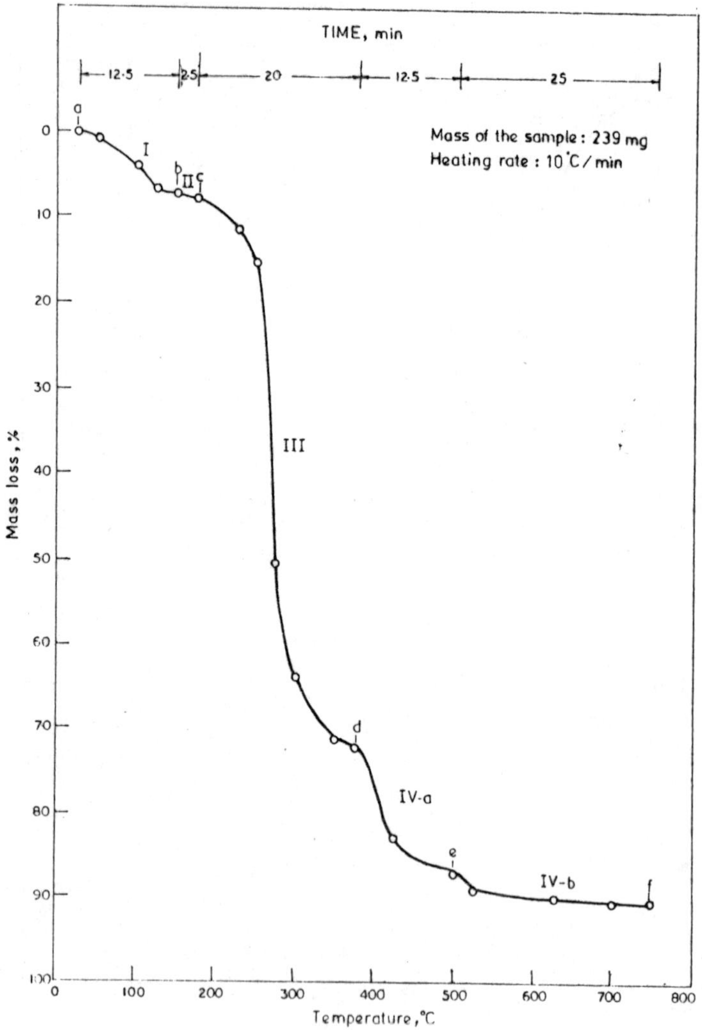

Fig. 14.2: TGA Curve of Ground and Acid Leached Straw

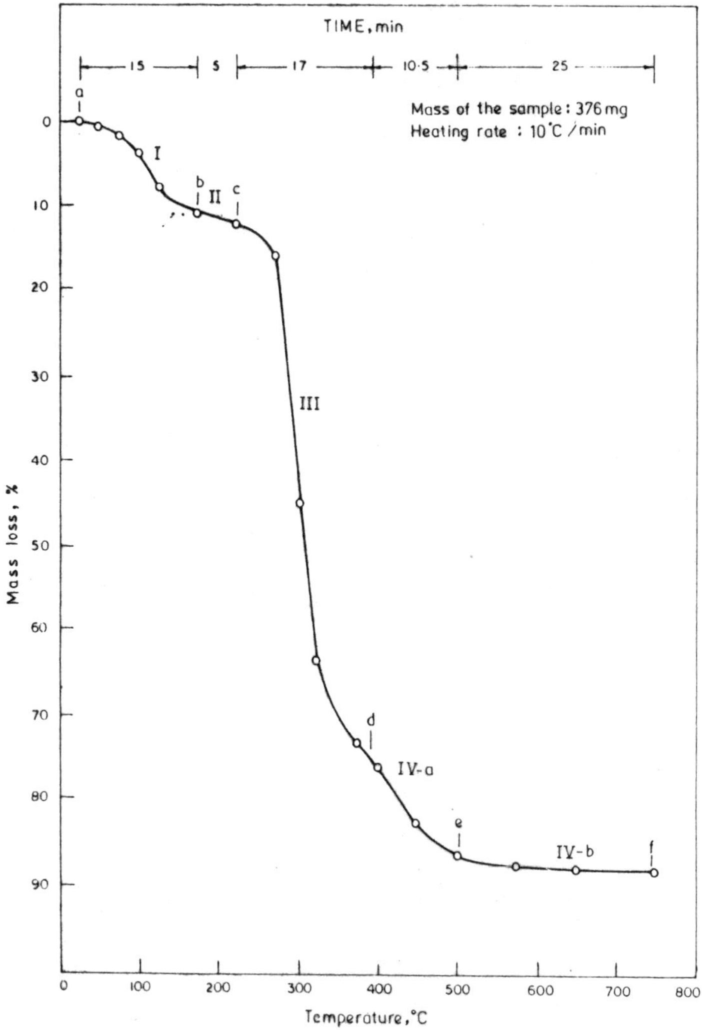

Fig. 14.3: TGA Curve of Ground and Raw Rice Husk

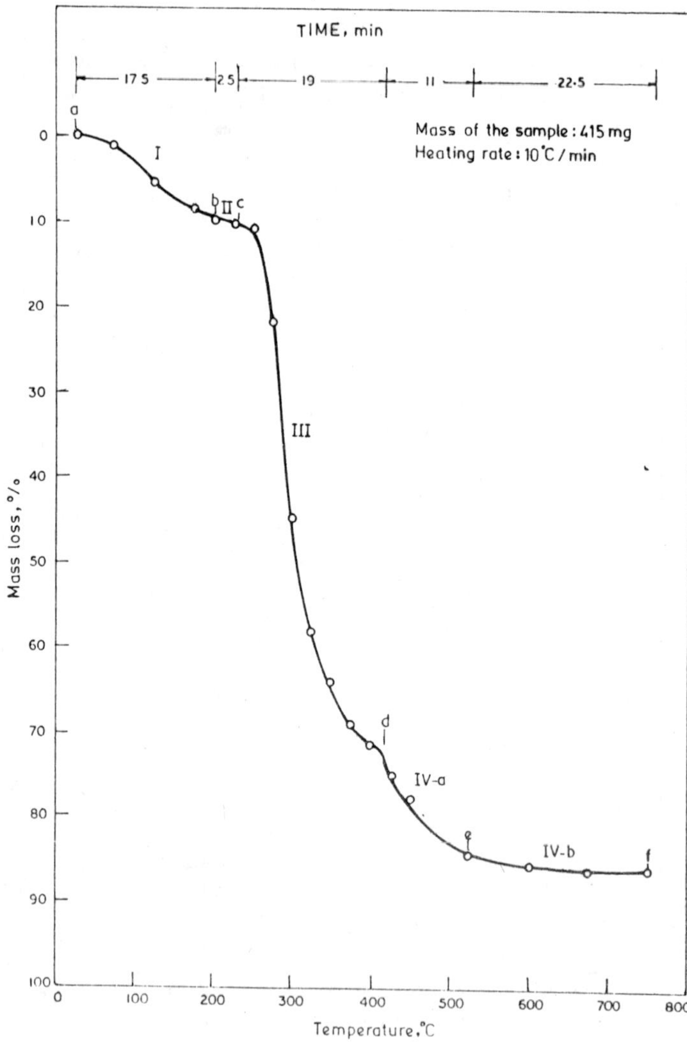

Fig. 14.4: TGA Curve of Ground and Acid Leached Rice Husk

reasonable period of combustion.

The acid treated straw and husk samples of 20 cm bed thickness subjected to 450° and 500°C furnace temperatures correspond to the complete combustion times of about 3 and 1.5 h and the maximum bed temperatures of 680° and 720°C for straw and 660° and 720°C for husk for the production of milky white ash. The combustion times required for 20 cm bed depth at 450° and 500°C were less for acid treated straw and husk compared to untreated samples.

Though the maximum bed temperature recorded for 20 cm bed thickness at 500°C was 720°C the diffractograms of both untreated and acid treated straw and husk did not show any sharp peak which revealed that the silica in the above white ash samples was amorphous in nature. As the bed temperature may go above 720°C, at higher bed depth and higher furnace temperature, a furnace temperature of 450°C and bed depth of 20 cm are suitable for the production of pure amorphous silica from 1N HCl treated straw and husk.

CONCLUSIONS

1. The thermal decomposition of raw and 1 (N) HCl treated rice-husk and straw in still air takes place in three distinct stages namely, (i) drying, (ii) devolatilization, and (iii) slow combustion of fixed carbon.
2. For the production of combustible gas along with the amorphous silica, both raw and acid treated straw and husk containing 6 to 7% (wb) moisture should be subjected to a furnace temperature of 450°C.
3. A combination of a furnace temperature of 450°C and a bed depth of 20 cm is recommended for the production of amorphous white silica from 1 (N) HCl acid treated straw and husk.

REFERENCES

1. Chakraverty, A., "Biotechnology and Other Alternative Technologies for Utilization of Biomass/Agricultural Wastes", Oxford and I.B.H., 1989.
2. Mishra, P., Chakraverty, A., and Banerjee, H.D., Thermochim, Acta, Amsterdum, 20 (1985) 4387-4391.

15

DEVELOPMENT OF 50 KW GROUNDNUT SHELL BASED GASIFIER SYSTEM FOR POWER GENERATION

—S.R. PATEL AND VIKAS KUMAR

In India, 7.5 million hectare area is put under groundnut cultivation and which yield 23.54 million ton of groundnut, out of which 7 million ton (30%) constitute groundnut shell[1]. As per MNES studies only 30% groundnut shell is utilised as boiler fuel in oil mills whose energy utilization efficiency is far from satisfactory and rendering 35% groundnut shell is available as surplus. The groundnut shell if gasified and used in oil mills for power generation can effectively meet substantial part of its electrical energy requirement. The major advantage of gasifying groundnut shell at oil mills is that it is easily available at oil mills so transportation cost of fuel can be saved.

It seems that there is a great potential for utilizing the groundnut shell for power generation through gasification route in oil mills.

DEVELOPMENT OF THE GASIFIER SYSTEM

In order to determine important parameters for designing a 50 kW groundnut shell based gasifier, a 3.7 kW laboratory scale, open-core, throatless, downdraft type gasifier system was tested on thermal mode.

Based on the data available from laboratory scale studies, a 50 kW groundnut shell based gasifier and a gas cooling and clean-up unit was designed, fabricated and installed at the Institute. The system consisted of an open-core, downdraft throatless gasifier alongwith pneumatic conveying system for continuous feeding of groundnut shell and a power operated rotating grate for continuous removal of ash and a gas cooling and clean-up unit. A schematic diagram of the gasifier system is given in Fig. 15.1. Details of the system are given below.

Gasifier

A throatless, open-core, down draft, groundnut shell based gasifier was designed. The designed capacity of gas production of the gasifier was 105 N cu.m. per hour. The gasifier was made into two parts (i) the upper part cylindrical hopper and (ii) the lower part ash pit.

The reactor consisted of two concentric cylinders. The inner cylinder was made of 10 mm thick mild steel sheet to protect any deformation due to high temperature in combustion zone and for longer life. It was 650 mm in diameter and 2000 mm in length. It worked as a reactor. The groundnut shell holding capacity of the reactor was about 50 kg. The outer cylinder was made of 3 mm thick mild steel sheet and it was of 750 mm in diameter and 2300 mm in length which acted as a heat exchanger and as a separation zone for coarse particles.

Feeding Unit

A pneumatic feeding system was provided to reduce manual labour involved in loading the gasifier. The feeding system consisted of an electric blower of 2 hp capacity and cyclone separator for separating air and groundnut shell.

Ash Removal System

For continuous operation of gasifier, a rotating grate was provided which removed gasifier residue i.e., ash and char from the reactor. The grate was made of 10 mm MS rod fixed equally at the spacing of 20 mm on the periphery of MS ring. The diameter of grate was 500 mm. The grate was connected with power operated shaft with bevel gear arrangement through which power was supplied through 1 hp DC motor alongwith control panel. In order to control the ash removal rate, necessary arrangements were made, such as to vary the distance between grate and bottom of inner cylinder of gasifier and

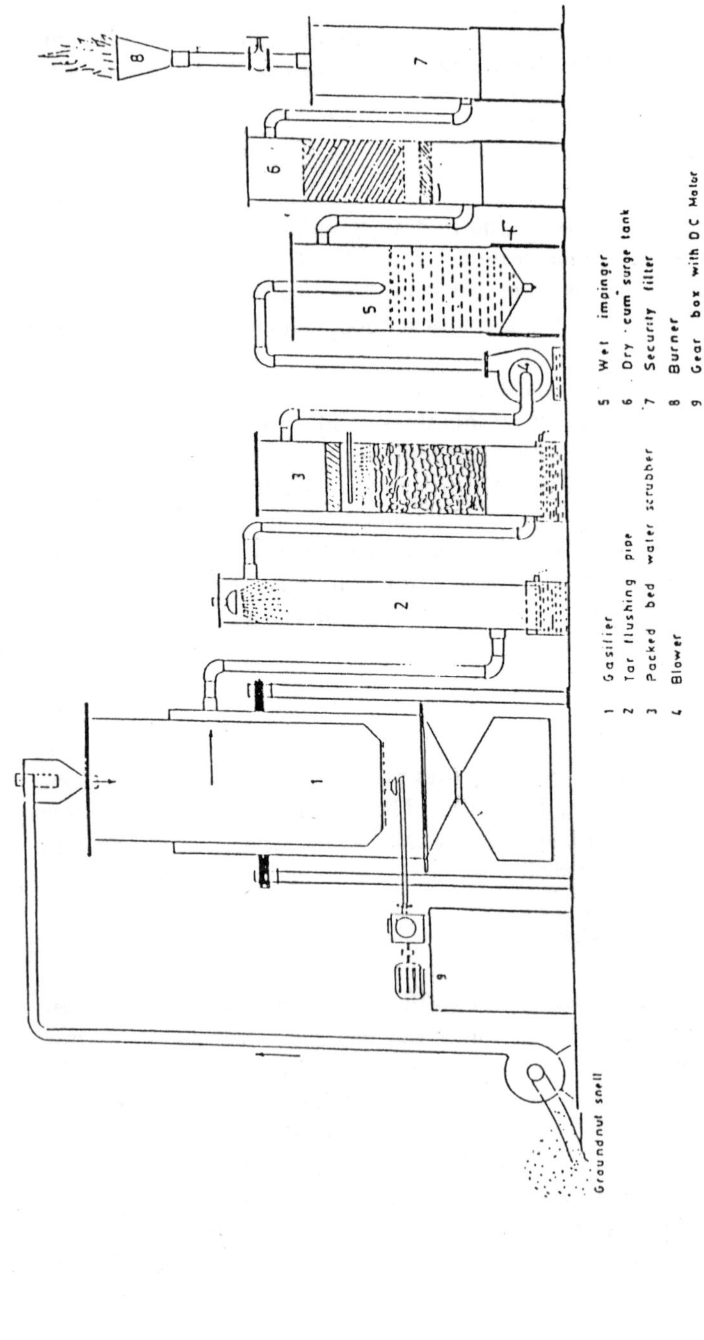

Fig. 15.1: Schematic Diagram of 50 kW Groundnut Shell Based Gasifier System under Thermal Mode

rotation of grate through control panel. The function of control panel was to vary the rpm of the DC motor. The ash was collected at the lower cone of outer reactor. The cone was connected with an ashpit. The holding capacity of groundnut shell char in ash pit was 15 kg. It was connected through 100 mm gate valve and flange system.

Gas Cooling and Clean-up Unit

A suitable gas cooling and clean-up unit was designed. The gas cooling and clean-up unit consisted of a tar flushing pipe, a packed bed water scrubber, a dry filter, a wet impinger, a surge tank-cum-dry filter and a security filter. The details of the system are as follows.

Tar Flushing Pipe

The tar flushing pipe was a vertical empty pipe and it was directly connected to the gas outlet of the gasifier. It was 3000 mm in length and having 200 mm diameter. Water was sprayed from the top and gas water contact was co-current. It removed coarse particles and reduced the temperature of the gas condensed. The water requirement was 800±50 liters/h.

Packed Bed Water Scrubber

A packed bed water scrubber was designed[2]. It was a M.S., cylinder of 1500 mm in length and 500 mm in diameter with gas outlet at the top and gas inlet at the bottom. Smooth pebbles of 35-40 mm size were used as packing material. The height of bed was 600 mm. The filter contained mist eliminator at the top of the scrubber which helped in condensing water vapour. The partly processed gas entered at the bottom and flown upward through distributed space of packing material. A water spray arrangement was provided to irrigate the bed. It effectively removed the soot particles and condensed the tar varpour from the gas.

Dry Filter

It was a M.S. cylinder having 500 mm diameter and 450 mm length alongwith mist eliminator. This mist eliminator helped in condensation of water and tar vapour.

Wet Impinger

The wet impinger was designed which worked by impinging producer gas through a nozzle on the water surface[3]. The diameter and the length of the wet impinger were 400 and 800 mm respectively whereas the nozzle diameter was 25 mm. The distance between water level and nozzle was kept 10 mm. The impinger needed only 65 liters of water for 25 hours of operation of the gasifier system. Thus it saved considerable amount of water also.

Surge tank-cum-dry-filter

It was of 600 mm in diameter and 1000 mm in height. It condensed moisture from the gas and supplied gas at a constant pressure to the engine. A mist eliminator was provided to condense water vapour.

Security Filter

It was of 400 mm in diameter and 1000 mm in length. The filter contained densely packed cotton waste. It removes any fine particles that may still remain in the gas and it absorbed moisture present in the gas.

PRELIMINARY TESTING

In order to evaluate the performance of the gasifier system, a test set-up was prepared. The test set-up consisted of a gasifier, a gas cooling and clean-up unit and a centrifugal electric blower of 2 hp capacity alongwith a producer gas burner. The system was operated on thermal mode for 50 hours and necessary arrangements for measurement of temperature, pressure, gas flow rate were made. During operation of the gasifier system, it was observed that performance of the gasifier was satisfactory but the following major problems were faced.

(i) In gasifier, dry rotating grate was not working satisfactorily because it was exposed to high temperature continuously, which resulted in deformation of shaft and its connections.

(ii) Total tar and particulate content in the gas after the gas cooling and clean-up unit was higher than the allowable unit.

To avoid these problems, the following modifications in the system were incorporated.

MODIFICATION

1. A wet type ash removal system was introduced. The ash pit consisted of a rectangular M.S. (Sheet of 3 mm) tank having dimensions of 1700 × 800 × 500 mm in which water was filled for cooling and sealing the reactor. The holding capacity of tank was 500 litres. The grate shaft made of 35 mm MS rod rotates in two ball bearings and passes through the ash pit at perpendicular axis. The shaft was directly coupled to a reduction gear box alongwith the electric D.C., motor and control panel. The distance between grate and bottom of gasifier reactor was kept at 250 mm. The ash removal rate was found to be 0.429 kg per rotation. The details of the rotating grate assembly are given in Fig. 15.2. The overall performance of the wet system was satisfactory.
2. In gas cooling and clean-up unit, a 2 hp electric blower was connected after dry filter for two purposes; (i) Suction of gas from the reactor and (ii) Condensation of tar vapour due to centrifuge action. For further reducing tar content in the gas, water was introduced into the impeller of electric blower which helped in better mixing of water and gas. Thus it helped in condensation of tar inside the blower and removal of soot particles and it worked as a hydro-tar-extractor-cum-blower. Necessary arrangements for continuous removal of waste water is shown in Fig. 15.3. Similarly a 0.25 hp electric blower was connected after wet impinger.

PERFORMANCE OF THE SYSTEM

The modified gasifier system consisted of an opencore gasifier and a gas cooling and clean-up unit. The gas cooling and clean-up unit consisted of a tar flushing pipe, a packed bed water scrubber, a dry filter, a 2 hp electric blower cum hydro-tar-extractor, a wet impinger, a 0.25 hp electric blower cum hydro-tar-extractor, a surge tank-cum-dry filter and a security filter. Figure 15.4 and 15.5 show schematic diagram and photograph of the gasifier system respectively. To evaluate the performance of the gasifier system, the system was

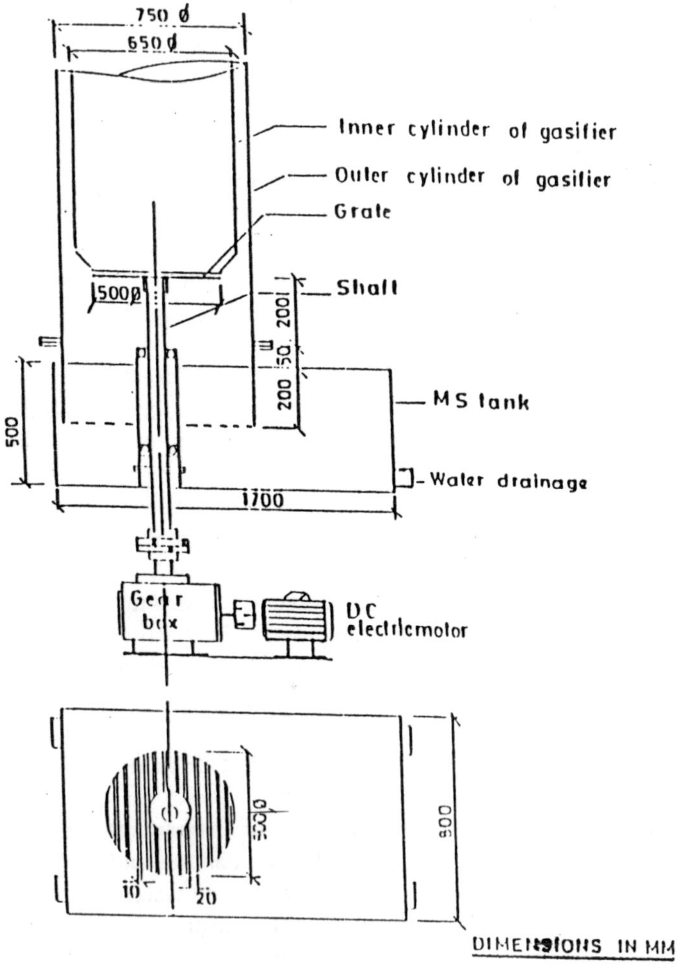

Fig. 15.2: Details of Rotating Grade Assembly

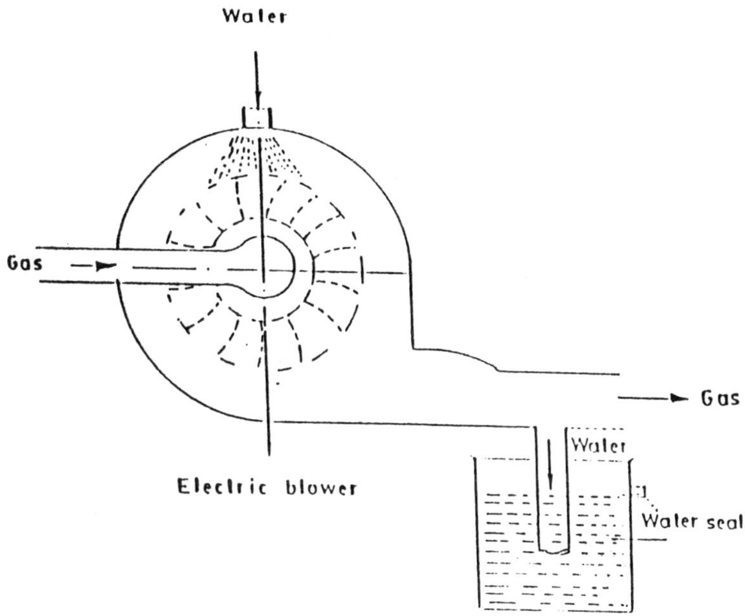

Fig. 15.3: Arrangement of Water Injection in Electric Blower and Continuous Removal of Waste Water

operated on thermal mode at different gas flow rates varying from 45 to 100 N cu. m per hour and groundnut shell consumption was in the range of 33.8 to 69.3 kg per hour. The average total tar and soot content in the gas cooling clean-up unit was in the range of 200 to 250 mg/N cu.m. Which was within satisfactorily limit for IC Engine application[4]. The overall performance was satisfactory.

CONCLUSIONS

1. A 50 kW groundnut shell based gasifier system was designed, fabricated and installed at the SPRERI. Based on the preliminary testing, necessary modifications were incorporated.
2. The modified gasifier system was operated on thermal mode for 250 hours without any technical problem. The gasifier was producing consistent quality of gas throughout the test run.
3. Tar and Particulate contents in the gas after the gas cooling

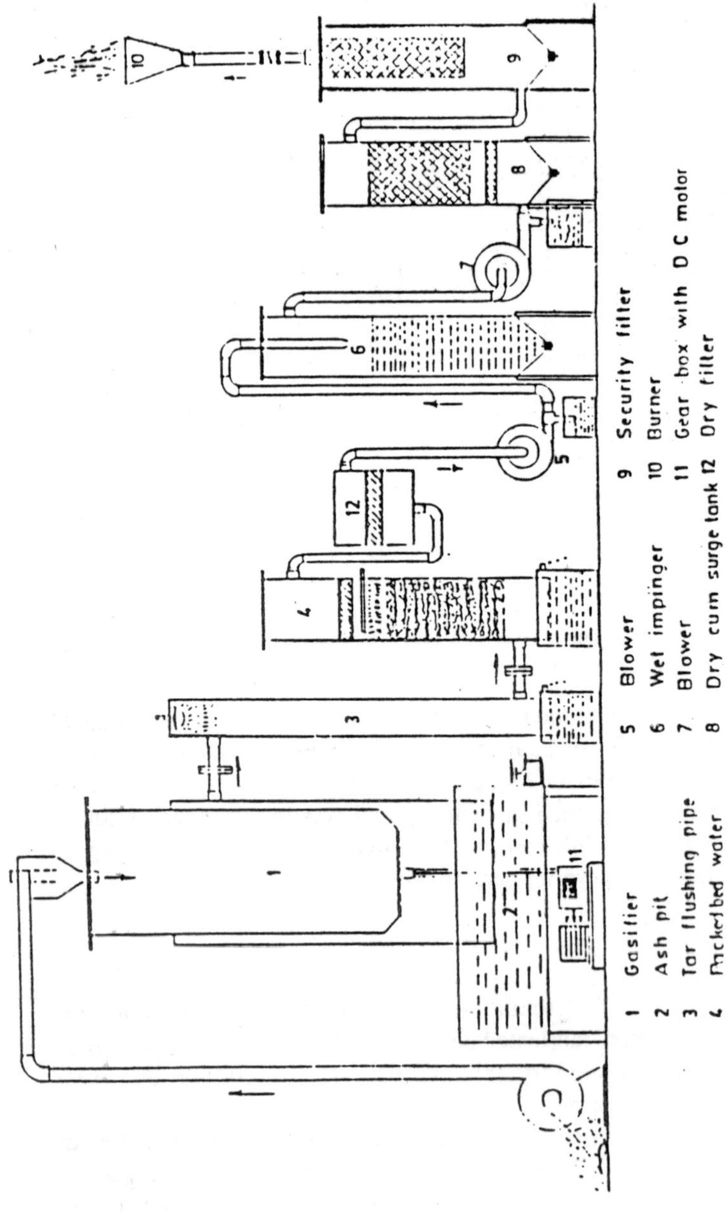

Fig. 15.4: Schematic Diagram of Modified 50 kW Groundnut Shell Based Gasifier System under Thermal Mode

Fig. 15.5: Photograph of Final 50 kW Groundnut Shell Based Gasifier System under Thermal Mode

and clean-up unit was in the range of 200 of 250 mg/cu.N.

ACKNOWLEDGEMENT

The authors are grateful to the Director and the management of SPRERI for providing the facilities to carry out the project work. They are also grateful to the Gujarat Energy Development Agency, Vadodara, for extending financial support.

NOTES

1. ICAR Hand book of Agriculture (1990), *Commercial Crops*, pp. 923.
2. Perry P.H. and Chilton, C.H. (1973). Packed Columns, Chemical Engineer's Hand Book Company, New York, 18. pp. 19-49.
3. Khummongkol, D. and Tangsathitkulchai, C., (1989). A model for tar removal efficiency from biomass produced gas impinging on a water scrubber', *Energy*, Vol. 14(3); pp. 113-121.
4. Anon, *Technical Leaflet of the TNO Institute for Road Vehicles*, Delft, Netherlands, 1984.

PART SEVEN

BIOMETHANATION

16

FLUIDIZED-BED ANAEROBIC DIGESTION-REACTOR DESIGN AND EXPERIMENTATION

—D. SUBHAKAR

In the past, the anaerobic contact processes (anaerobic activated sludge) has been applied to the treatment of wastes of several processes[1]. Lack of fundamental knowledge of the processes and its limitations have probably limited more widespread usage of this process. The anaerobic filter for the treatment of concentrated soluble organic wastewaters was found to be able to treat dilute wastes in a short period of time at ambient temperatures. However, this process is hampered by clogging and inefficient contact of the micro-organisms and wastewater. Under aerobic conditions, static attached film had twice the organic removal capacity of a suspended microbial system under comparable conditions. In a subsequent study, Jewell[2] proposed the attached film expanded bed process and demonstrated that it is possible to utilise this concept using anaerobic film. This concept has been applied to anaerobic systems and found to be highly efficient at short hydraulic times. Pilot plants containing large scale fluidized bed bioreactors as advances treatment processes for nitrification and denitrification of industrial wastewater have been successfully employed[3]. A similar process has been employed to remove nearly 80% of BOD at high organic loading rates with a simultaneous generation of methane and stabilised high strength wastes including from dairy, chemical, food processing, soft drink bottling and heat transfer liquids. This system is a net energy producer in most cases and normally shows a payback period of five years. The advantages of this process include,

1. greater surface area available for growth per unit of reactor volume
2. very small head loss
3. no danger of clogging
4. easier carrier removal procedure
5. diminished labour cost
6. facilitation for automatic control
7. greater consistency in reaction conditions and hence greater consistency in quality of the product.

The following factors constitute the effectiveness of fluidized bed process:

1. Maximum contact between the liquid and fine grained media.
2. Minimal liquid film diffusional resistance attributable to the particle motion and liquid velocities.
3. Problems of channelling, plugging and gas hold-up commonly encountered in packed beds are avoided.

In recent years, great strides have been made in understanding and modelling biofilm kinetics, most of the progress has been achieved by idealising the process of substrate utilisation, molecular diffusion, and mass transport as simultaneous differential equations for a homogeneous layer of bacteria. Experimental work with artificially produced films has verified that these three concepts the essentials of biofilm kinetics of substrate utilisation[4]. On the basis of this conceptual model, a theoretical study is initiated to accomplish the following objectives:

1. To analyse and modify existing models to obtain a theoretical model, of substrate utilisation kinetics within biofilms applicable for fluidized bed anaerobic digestion process.
2. To use this model as a tool for understanding the important factors controlling substrate utilisation by biofilms.
3. To test this model by a detailed experimentation to find out the inter-relationship between various parameters of the process.

FLUIDIZED BED ANAEROBIC DIGESTION PROCESS

Theory of Fluidization

A fluid flowing at low velocities through a porous bed of small solid particles may not cause the particles to move. However, if the fluid velocity is increased further, a point is eventually reached at which the particles no longer remain stable, but 'fluidize' under the action of the fluid. At that point the particles are supported by the fluid. When the fluid velocity through a bed of solids cause all of the particles to be entrained in the fluid and to be carried along with it, this is called 'fluidization'. This is used primarily to transport solids in a gas stream in most industries.

Parameters of Fluidization

To model the kinetics or the dynamics of these biologically active fluidized beds, it is important to establish the relationship between superficial liquid velocity and expanded bed voidage. Ngian and Martin have experimentally verified that

$$u/v = \varepsilon_u^n \quad \ldots\ (1)$$

For small size particles, the values of v and n are given by the relationship

$$v/u_t = 10^{-dm/D} \quad \ldots\ (2)$$

$n = 4.65 + 20\, d_m/D$ for $Re<0.2$

$n = (4.4 + 18\, d_m/D)\, Re_t^{-0.03}$ for $0.2<Re_t<1$

$n = (4.4 + 18\, d_m/D)\, Re_t^{-0.1}$ for $1.0<Re_t<200$

$n = 4.4\, Re_t^{-0.1}$ for $200<Re_t<500$

$n = 2.4$

The terminal or free fall velocity of the particles is estimated using[5]

$$u_t = \frac{g(\rho_p - \rho_f)^2\, d_p^2}{18\, \mu^2} \quad \text{for } Re_t < 0.4 \quad \ldots\ (3)$$

$$u_t = \left[\frac{4g(\rho_p - \rho_f)^2\, d_p^2}{225\, \rho_l \mu^2}\right]^{\frac{1}{3}} \quad \text{for } 0.4 < Re_t < 500$$

$$u_t = \left[\frac{3.1g(\rho_p - \rho_f)^2\, d_p}{225\, \rho_t}\right]^{\frac{1}{2}} \quad \text{for } 500 < Re_t < 200000$$

The relationship between liquid superficial velocity and porosity of a biologically active bed given by Eqn. (1) can be represented as,

$$ln\ u = ln\ v + n\ ln\ \varepsilon_u \qquad .... \qquad (4)$$

Using this plot, ε_{mf} minimum porosity for fluidization can be calculated.

Since the mass of the material of expanded and unexpanded bed remains constant, the height of the bed can be found using

$$LA\ (1- \xi_o)\ \rho_p = L_u\ A(1- \varepsilon_u)\ \rho_p \qquad .... \qquad (5)$$

Assuming the cross-sectional area of the bed a constant,

$$L_u = \frac{L(1-\xi_o)}{(1-\varepsilon_u)} \qquad .... \qquad (6)$$

The pressure drop is calculated using the Ergun equation,

$$\frac{\Delta P g_c}{L_{mf}} = 150 \left[\frac{(1-\varepsilon_{mf})^2}{\varepsilon^3_{mf}}\right] \frac{u\mu_l}{(\phi_s d_p)^2} + 1.75 \frac{(1-\varepsilon_{mf})}{\varepsilon_{mf}^{\ 3}} \frac{u^2\rho_l}{(\phi_s d_p)} \qquad .... \qquad (7)$$

For smaller particles of smaller specific weights, the superficial terminal velocity at minimum fluidization is given as,

$$u_{mf} = \frac{(\phi_s\, d_p)^2}{150} \frac{(\rho_p - \rho_l)}{\mu} \frac{g\varepsilon^3_{mf}}{(1-\varepsilon_{mf})} \qquad \text{for } Re_u < 20 \qquad ..... \qquad (8)$$

When ε_{mf} and ρ_s are unknown, the following modified equation can be used.

$$\frac{d_p u_{mf} \rho_l}{\mu_l} = \left[(33.7)^2 + \frac{0.0408 d_p^3\ \rho_l g\ (\rho_p - \rho_l)}{\mu^2_l}\right]^{1/2} - 33.7 \qquad ..... \qquad (9)$$

At the optimum fluidization point, the limiting superficial velocity should be equal to the terminal settling velocity and hence,

$$u_t = u_l = \frac{g d_p^2 (\rho_p - \rho_l)}{18\mu} \qquad .... \qquad (10)$$

The minimum porosity for fluidization can be approximated for particles smaller than 500 microns and larger than 50 microns by

$$\varepsilon_{mf} = 1 - 0.356\ (\log d_p - 1) \qquad .... \qquad (11)$$

Choice of Distributor

The quality of fluidization is strongly influenced by the type of distributor used. In designing a distributor, the available pressure drop is first determined and according to it, the type of distributor is selected. If the distributor have to pass a large amount of solids entrained by inlet fluid, perforated plates are recommended. Among various types of grids and distributors, perforated plate or wire mesh is commonly used for laboratory scale operations. Nozzles are used to prevent solids from falling through the distributor. The pressure drop across the distributor generally should not exceed 10% of the pressure drop across the bed.

Effect of Parameters on Digestion

From the studies made, it can be concluded that the following parameters influence the fluidized bed anaerobic digestion process. They are :

1. HRT
2. Temperature
3. Organic volumetric loading rate
4. Influent substrate concentration
5. Initial bed height or amount of clear media used
6. Expanded bed height
7. Size of the media
8. Biofilm thickness
9. Reactor area perpendicular to flow
10. Superficial flow velocity
11. Biofilm moisture content
12. Factors affecting quality of fluidization like design of distributor, baffles, etc.

KINETIC MODEL

Monod kinetic model

Monod in his experiments have shown that the effect of a limiting substrate can be defined as

$$\mu = \frac{\mu_m S}{K_s + S} \quad .... \quad (12)$$

Since the bacteria increase in proportion to their mass in the log-growth phase, $r_g = \mu_x$

$$= \frac{\mu_m xS}{K_s + S} \quad .... \quad (13)$$

The rate of substrate utilisation and rate of substrate growth is related as

$$r_g = -Y r_{su} \quad .... \quad (14)$$

Hence the rate of substrate utilisation is

$$r_{su} = \frac{-\mu_m xS}{Y(K_s + S)} \quad .... \quad (15)$$

The term μ_m/Y is often replaced by the term k, defined as the maximum rate of substrate utilisation per unit mass of micro-organisms. Hence

$$\frac{dS}{dt} = r_{su} = \frac{-kxS}{K_s + S} \quad .... \quad (16)$$

Kinetic Model for FBAD Process

The biofilm is assumed to be attached to a flat plate with infinite length and width. Also, it is assumed that the rate of reaction is limited by a single substrate S. Molecular diffusion, which is the only source of mass transport within the biofilm is related to the substrate concentration.

By Fick's law,

$$\frac{dS}{dt} = D\frac{d^2 S}{dx^2} = -\frac{kXS}{K_s + S} \quad .... \quad (17)$$

Assuming that the substrate concentration is very low compared with K_s ($K_s >> S$),

$$D\frac{d^2 S}{dX^2} = -KS \quad .... \quad (18)$$

Where $K = \frac{kX}{K_s}$

The boundary conditions for Eqn. (18) are :

$$\text{At } x = \delta, \frac{ds}{dx} = 0 \quad .... \quad (19)$$

$$\text{At } x = 0, S = S_0 \quad .... \quad (20)$$

For a continuous reactor, the solution is obtained as

$$S_b = \left[\frac{S_o}{1 + ct}\right] \left[\frac{S_o}{1 + E_u t}\right]$$ where

$$c = \frac{DA_p}{V} \sqrt{\frac{K}{D}}$$

Model Simulation

The model is simulated with a reactor of volume 500 cc, having 160 gm of 0.05 cm support media (2.79 gm/cc). The substrate concentration is 200 mg/lt and S_b is 78.3 mg/lt and time as one hour. The factor $D\sqrt{\frac{K}{D}} = 0.112928$. The experiments conducted have given a close agreement with the results obtained from this model.

Experimental Procedure

A laboratory scale anaerobic digester is used to study the process of fluidized bed anaerobic digestion. The reactor is made of perspex, and of 9 cm dia. and 90 cm height. The base is constructed to disperse the waste uniformly across the bottom by using stones of 1-2 cm dia. and this helped to improve the quality of fluidization. After many practical design, fabrication and experimentations, a distributor with nozzles having side holes is used for the final experiments. A slurry pump is used to circulate the natural waste (chicken waste). Bed is allowed to expand 100%. Other details of the gas and particle collectors, media preparation, etc., are explained in detail elsewhere.[6] The ranges of parameters attempted are :

Feed used	: Chicken waste
Support media	: Sand
Influent substrate concentration	: 1-5 gm/lt
pH	: 7
HRT	: 6-12 hrs
Size of sand particles	: 0.1-1.0 mm
Superficial fluid velocity inside digester	: 0.4-0.8 lt/min.
Unexpanded bed height	: 5-20 cm

Expanded bed height	: 5-40 cm
Nozzle diameter	: 1-3 mm
Organic loading rate	: 0.5-40 kg COD/m^3/day

Keeping all parameters constant and varying only one within the range given above, variation in gas production is observed. The methane content is estimated using Arsat apparatus. Other kinetic parameters could only be estimated using the literature available. Though methane is produced in the initial stages, prolonged experimentation made the film detach from the support particles and floated out. It is suggested that further experimentation with immobilisation techniques to fix the biofilm on the media is to be undertaken to make the fluidized bed reactor work continuously for longer durations continuously.

CONCLUSIONS

Fluidized bed anaerobic digestion is one of the recent developments is waste stabilisation. This process is suitable for the treatment of low strength organic wastes anaerobically, at reduced temperatures, at short retention times and at higher organic loading rates. A mathematical model using Monod kinetics is suggested for the fluidised bed anaerobic digestion and an analytical solution is obtained to predict the substrate utilisation rate. An experimental system is used to simulate the model parameters and the results show a close agreement. A distributed side nozzle type of distributor is found to give the required fluidisation than other types of distributors tried. Continuous gas production could be achieved if the biofilm is made to stick to the surface using immobilisation techniques.

ACKNOWLEDGEMENT

This study was carried out at AIT, Bangkok under the guidance of Prof. S.C. Bhattacharya.

Nomenclature

A_p Surface area of particle, cm^2
D Bed diameter, cm

D	Molecular diffusivity of the substrate into the biofilm, L^2/time
d_m	Diameter of the media, cm
d_p	Diameter of a sphere having the volume of the particle, cm
g	Acceleration due to gravity, cm/s
g_c	Conversion factor
K_s	Half-velocity coefficient, mass/unit volume
k	Minimum substrate utilisation rate per unit mass of micro organisms
L	Unexpanded bed height
L_u	Expanded bed height under fluid velocity u, cm
L_{mf}	Bed height at minimum fluidization conditions, cm
ΔP	Pressure drop across the bed, atm.
r_g	Rate of bacterial growth
r_{su}	Rate of substrate utilisation
r_m	Radius of the support media
S	Concentration of growth limiting substrate in solution, gm/lt
S_b	Bulk liquid substrate concentration, gm/lt
S_o	Influent substrate concentration, gm/lt
t	Time, s
u	Superficial liquid velocity, cm/s
u_{mf}	Minimum fluidization liquid velocity, cm/s
u_t	Terminal velocity, cm/s
u_l	Limiting superficial velocity, cm/s
V	Volume of digester, cm^3
x	Distance along the axis normal to biofilm, cm
X	Concentration of micro-organism, mass/unit volume
Y	Minimum yield coefficient, mass/mass
ξ_o	Unexpanded bed porosity
ε_u	Bed voidage at fluid velocity u
ε_{mf}	Bed porosity at minimum fluidization conditions
ρ	Density
μ	Viscosity
ϕ_s	Sphericity of a particle
μ_m	Minimum specific growth rate, time.
δ	Biofilm thickness

NOTES

1. McCarty, P.L. (1964). Anaerobic waste fundamental, Part I, Chemistry and microbiology, *Public Works*, Sep., pp. 107-112.

2. Switzenbaun, M.S. and Jewell, W.J. (1980). Anaerobic attached film expanded bed reactor treatment, *JWPCF*, 52 (7), 1953-1965.
3. Walker, J.F. *et al.* (1981). Biotechnology and Bioengineering Symposium No. 11, pp. 415-427.
4. Williamson, K. and McCarty, L. (1976). A model of substrate utilisation by bacterial films, *JWPCF*, 48 (1), pp. 9-24.
5. Kunni, D.K. and Devenspler, D. (1969). Fluidisation Engineering, Wiley, New York.
6. Subhakar, D. (1983). Fluidized-bed anaerobic digestion, Special study report, Asian Institute of Technology, Bangkok, Thailand.

17

PERFORMANCE OF PILOT BIPHASIC SYSTEM FOR THE GENERATION OF METHANE FROM AGRICULTURAL RESIDUES

—G.P. NAGORI AND N. SAHA

During recent years, biogas technology has emerged as an established technology for the generation of methane, a renewable and non-polluting source of energy. The technology provides multiple benefits to the society as well as to individual beneficiaries. This includes environmental protection, women's welfare, availability of fertilizer and rural development. Almost all rural sectors, in addition to dung have large supply of other biomass resources like crop residues and terrestrial as well as aquatic weeds. Some of these biomass resources such as banana stems, water hyacinth, eucalyptus leaves, kitchen waste and composite agricultural waste have successfully been utilized for biogas generation[1,2,3,4].

The biogas production is a complex biological process which converts biomass to methane and carbondioxide in absence of oxygen.

In the conventional single stage digestion system, the potential is not fully exploited due to inherent limitations of the traditional design. This is because, the set of micro-organisms responsible for hydrolysis, acid formation and for methane production are quite different from each other. They require altogether different environment to flourish. In biphasic system these two groups of organisms can be separated using two separate reactors. Both the reactors can be controlled at different conditions suitable to these sets of organisms which help in reducing the overall retention time also lowering of the system costs. Biphasic process has been successfully applied for feedstocks like water hyacinth[5], solid pineapple waste[6], crop and agricultural residues, vegetable waste in the laboratory scale. The feasibility of biphasic process of biomethanation of agricultural residues had also been studied at SPRERI using waterhyacinth and banana stems[4,7,8]. Based on the data obtained a pilot biphasic system was designed for the use of agricultural residues and installed on the Institute premises for preliminary studies. Present paper deals with the performance of pilot biphasic system developed by the SPRERI.

Materials and Methods

The pilot biphasic system consists of an acid reactor having effective volume of 0.5 cu.m. and a methane reactor having an effective volume of 1 cu.m.

Acid Reactor

Acid reactor is an open batch rectangular tank having a working volume of 500 litres. It was fabricated from mild steel plate of 3 mm thickness. Dimensions of the tank are 975 mm × 760 mm × 935 mm. The reactor has a conical hopper bottom from where the percolate is withdrawn through the outlet. The tank is supported on four 25 × 25 × 5 mm iron angles welded at four corners of the tank.

The acid reactor was charged everyday with 20 kg of agricultural residues and 160 litres of water initially. Fresh biogas slurry had been used for initial culture development. The feedstocks were chopped into 2.5 cm pieces using a chaff cutter and then fed into the acid reactor. The contents of the acid reactor were mixed thoroughly four times a day—twice in the morning and twice in the evening for 15 minutes. The liquid effluent of this reactor was taken out from the outlet provided at the bottom and filtered through a filter cloth and used as a substrate from methane reactor. The pH of the effluent of the acid reactor was measured everyday using a digital pH meter and the VFA and COD was analysed twice in a week using standard methods[9]. The feedstock to the acid reactor were also analysed for their total solids and volatile solids contents once in a fortnight using standard methods[9].

Methane Reactor

An anaerobic filter was used for methane reactor. The reactor was constructed of prefabricated reinforced concrete pipe having an internal diameter of 1.2 m and height of 2.5 m. To support the packing media and to distribute the flow of influent, a dispersion plate of 6 mm thickness with 10 mm diameter holes is provided about 22 cm above the bottom of the reactor. Brickbats having a mean diameter of 5 cm were used as packing media to support the growth of bacterial. For creating anaerobic conditions and for collecting the gas, a mild steel gas holder was provided at the top of the reactor. The gas holder has internal diameter of 1.15 m and height of 0.3 m. The diameter of the roof of the gas holder is kept 1.3 m which is equal to the outer diameter of methane reactor, so that the gas holder can rest on the walls of the methane reactor when there is no gas in it.

Gas outlet is provided at the centre of the gas holder which is connected to gas flow meter using flexible rubber pipe. The effluent outlet is provided near the top of the reactor and inlet pipe of 25 mm diameter was at the centre of the bottom. A diaphragm pump was installed near the reactor and inlet and outlet of the reactor were connected using 15 mm galvanized iron pipes, fittings and valves. Fig. 17.1 shows the flow diagram of prototype biphasic system.

The reactor was fed with 70 litres of screened digested slurry using diaphragm pump at a flow of 64 litres/hour. The reactor contents were recycled twice a day for 30 minutes keeping the same flow rate to enhance the bacterial growth and to avoid clogging. Gas produced was collected in inverted glass bottles over water and measured everyday. When the gas production from culture reached steady state the feeding of acid rich effluent from acid reactor into the methane reactor was started. Initially, the reactor was operated at 15 days retention time which was gradually brought down to 8 days and steady state data were collected at this HRT. During steady state influent and effluent pH and gas production were measured everyday. The influent and effluent VFA and COD were analysed once in a week using standard methods[9]. The gas was analysed once in a week using SIGMA make gas chromatograph on TCD using porapak Q column. Nitrogen was used as carrier gas and the flow was maintained at 40 ml/minute.

Results and Discussions

The Pilot biphasic system consists of acid reactor and methane

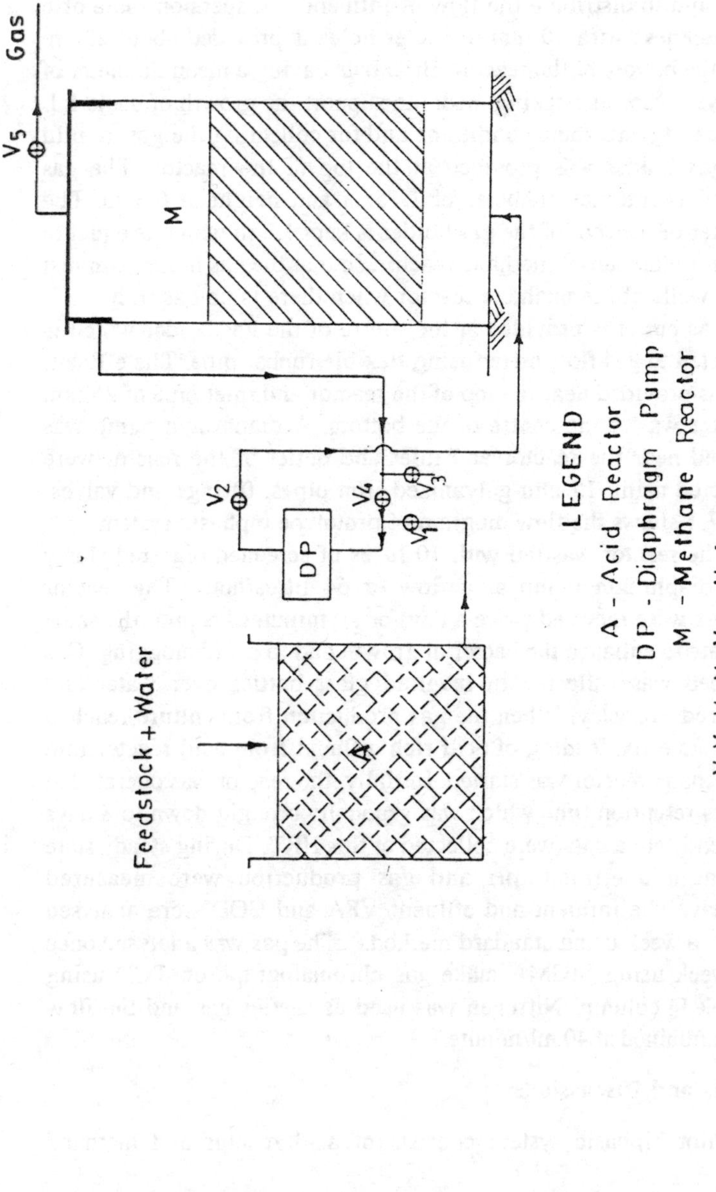

Fig. 17.1: Flow Diagram of Prototype Biphase System

reactor. The system was commissioned in the institute premises and operated using agricultural residues like green banana stems and waterhyacinth. The performance data for acid reactor for both the feedstocks used is shown in Table 17.1.

TABLE 17.1: **Performance of Pilot Acid Reactor**

Feed stock	Green banana stems	Green water hyacinth
HRT (days)	3	3
Average ambient temperature (degree celsius)	21-26	22-33
Feed Rate		
(i) Feedstock (kg/day)	20	20
(ii) Water (*l*/day)	160	160
Feed characteristics		
(i) Total solids (% of wet wt)	5.21-5.75	7.25-7.94
(ii) Volatile solids (% of dry wt)	79.35-80.05	84.60-87.07
Effluent Characteristics		
(i) pH	6.75-6.90	6.40-6.65
(ii) VFA (mg/*l*)	1,515-1,635	1,550-1,820
(iii) COD (mg/*l*)	8,420-8,710	8,515-9,525

It is observed from Table 17.1 that the effluent from acid reactor was slightly acidic for both the feedstocks. The VFA and COD for banana stems were in the range of 1,515-1,635 mg/*l* and 8,420-8,710 mg/*l* respectively whereas the VFA and COD for water hyacinth were in the range of 1,550-1,820 mg/*l* and 8,515-9,525 mg/*l* respectively. The higher VFA and COD in the effluent from acid reactor for waterhyacinth were because of the higher ambient temperatures during the period of operation of acid reactor on waterhyacinth. While the average ambient temperature was in the range of 21-26 degrees celsius when banana stems were used as feedstock. At the temperature range of 29-33 degree celsius the bacterial activity/activities enhance. As a result the concentration of VFA and COD in the effluent of acid reactor was shown high.

The performance data for methane reactor at various HRT is shown in Table 17.2. It is observed from Table 17.2 that the biogas production was found to increase with decrease in HRT. This was obvious as the quantity of influent was increased with a decrease in HRT. Hence, more organic matter was available for decomposition. The maximum gas production recorded was 320 litres/day at 8 days HRT. It was also observed that the biogas yield was 0.25 litres/g COD added for both 15 days and 12 days HRT whereas it was increased to 0.28 litres/g COD added for 10 days HRT and was maximum of 0.30 litres/g COD added for 8 days HRT. The increase in biogas yield for run at 10 days and 8 days HRTs was attributed to increase in average ambient temperatures. The temperature were 23 and 24 degree Celsius for the run at 15 and 12 days HRT respectively and they were 30 and 32 degrees celsius for the run at 10 and 8 days

TABLE 17.2: **Performance of Pilot Methane Reactor**

HRT(days)	15	12	10	8
Average ambient termperature (degree celsius)	23	24	30	33
Volume of influent (litres/day)	64	60	96	120
pH				
(i) Influent	6.84	6.71	6.64	6.79
(ii) Effluent	7.15	7.20	7.21	7.15
VFA (mg/*l*)				
(i) Influent	1,585	1,520	1,645	1,670
(ii) Effluent	340	330	345	330
(iii) Reduction (%)	78.5	78.3	79.0	80.2
COD mg/l				
(i) Influent	8,590	8,550	8,875	8,920
(ii) Effluent	1,795	1,840	1,830	1,785
(iii) Reduction (%)	79.1	76.5	79.3	80.6
Biogas production (litres/day)	140	175	235	320
Biogas yield (litres/g COD added)	0.25	0.25	0.28	0.30
Methane in Biogas (%)	60	61	81	82

HRTs respectively. As the average ambient temperatures of 30-32 degrees celsius are optimum for bacterial growth, it resulted in higher biogas yield. It can also be seen that the VFA and COD reduction was also increased at these HRTs owing to increased bacterial activities. The average VFA and COD in the influent to methane reactor were in the range of 1,520-1,670 mg/*l* and 8,550-8,920 mg/*l* respectively whereas VFA and COD in the effluent from methane reactor were in the range of 330-345 mg/*l* and 1,765-1,840 mg/*l* respectively. This shows a reduction of 78 and 80%.

It was also observed that though the pH of the influent was acidic in nature the pH of the effluent remained well on alkaline side. This shows the establishment of a strong buffer system in the fermenting mass which readily overcome the acidity of the influent without disturbing the pH of fermenting mass. The methane in biogas was in between 80-82%.

Conclusion

The biphasic biogas reactor system which has been developed is simple in design and in operation and accepts both agricultural residues (i.e. green water hyacinth and green banana stems). No major operational problems were encountered during the studies. There was no clogging of the filter. The recirculation flow also served to completely mix the liquid within the filter.

The acid reactor was fed everyday with 20 kg of agricultural residues and 160 litres of water. However the removal of effluent caused some problem as it contained high suspended solids. Secondly, the mixing in acid reactor was done manually which was laborious one.

The performance data indicate that it is possible to reduce the retention time further to 6.5 or even 5 days as the lowering of HRT from 15 days to 8 days did not affect the reduction in VFA and COD and the pH of the fermenting liquid.

Acknowledgement

The authors gratefully acknowledge the support and facilities provided by the Director and Management of SPRERI for the project. They are thankful to the Ministry of Non-conventional Energy Sources, New Delhi for extending the financial support.

NOTES

1. Nagori, G.P. and Patel B.D. Performance of a 5 cu.m. biogas plant using banana stem, SESI-5, Vol. 2, (1&2), 1-11, 1988.
2. Nagori, G.P. and Rao, C.S. Biogas Manure Plants based on Agricultural Residues, Published by Sardar Patel Renewable Energy Research Institute, Vallabh Vidyanagar, 1988.
3. Nagori, G.P. and Rao, C.S. "Development of a Biogas plant for Kitchen Waste—a case study, Presented at SPRERI—IIM Workshop on R&D Management, Vallabh Vidyanagar, Sep. 24, 1990.
4. Nagori, G.P. and Rao, C.S., "Two Phase Anaerobic Digestion of Banana Stem" Paper presented at the national Solar Energy Convention-92, IIT, Delhi. Dec., 28-30, 1992.
5. Saraswat, N. and Khanna, P. Methane recovery from water hyacinth through anaerobic activated Sludge process, Biotechnology and Bioengineering, Vol. XXXVIII, 240-246, 1986.
6. Tanticharoen, M. Supajanya, N. and Bhumiratana, S. Two stage anaerobic digestion of solid pineapple waste, Proc. 2nd Asian Workshop on Biogas Technology, 301-311, 1984.
7. Final Report of Laboratory Studies for assessing the feasibility of using Biphasic process for Biomethanation of Agricultural Waste, Submitted to Department of Non-conventional Energy Sources, Ministry of Energy, Government of India, New Delhi, 1991.
8. Nagori, G.P. and C.S. Rao. "Biphasic Anaerobic Fermentation of water hyacinth, Role of Renewable Energy in Energy Policy : Proceeding of the 17th National Solar Energy Convention, 218-222, 1994.
9. Standard Methods for the Examination of water and waste water, 13th Edition, American Public Health Association, Year Book Publication, Chicago, 1971.

18

CONVERSION AND UTILIZATION OF AGRO-INDUSTRIAL BY-PRODUCTS AND WASTES

—C.B. SINGH, H.K. SINGH AND BANWARI LAL

Generation of wastes is closely related with population growth of human and animals due to speeding up of industrialization, urbanization and commercialization. Both are the opposite sides of the same coin. More than 20 cities in India belong over 1 million of human population. Due to which, the wastes in the form of solid and liquid are growing exponentially for which a very strong integrated waste management system is to be planned, organised and controlled. Out of many conventional and non-conventional sources, later is by far best alternative not only for remote areas but also for whole country.

On account of commencement of advance technology development it has been proved that there is nothing in the world to which we can say waste material. Every piece of crop residues, wastes of human and animals and by products from industries can be used more efficiently after suitable conversion than that of main material. Conversion of these products in the another required shape is useful not only for minimization and disposal of wastes but also to keep clean the cities and environment free from pollution. But in each case the alternatives must be evaluated to assess their likely impact on the environment, the economy and society/human health.

Power Production through Gasification of Farm Residues

According to an estimate, about 200 million tonnes of agricultural residues are produced every year in India. Crop residues and by products can be used as a source of heating either directly or through by-products (after conversion) because gross calorific values of these wastes are about 15-20 MJ/kg which is almost equal to that of 16 MJ/kg of diesel and 20 MJ/kg of dried wood. The crop residues can be converted into energy and power using thermo-chemical method (gasification) to produce low or intermediate calorific values producer gas with the help of gasifier (Singhal and Solanki, 1995).

However, coal powder and wood chips have been two principal feed stocks in the gasifers, but agricultural residues like real husk, groundnut shells, cotton and arhar sticks and maize cobs can easily be used to generate the producer gas. General composition of the produce gas is as below:

CO	:	20-22%
H_2	:	15-18%
CO_2	:	9-11%
N_2	:	50-53%
CH_4	:	2-4%

Management of Municipal Solid Wastes

The generation of municipal solid wastes in our cities has increased several fold during last few years due to accelerated increase in population. The urban share of India's total population is expected to grow from 23% to 32% over the period 1981 to 2001. The urban population at the end of this century is estimated around 33 crore (Kala and Khan, 1994). According to a sample survey, about 350 to 400 gms of per capita per day solid waste is generated in an average Indian town. In major metropolitan towns, it may rise to approximately 400 to 425 gms. While in Delhi, it is around 800 to 850 gms per capita per day. Solid wastes generation in major cities of India has been shown in Table 18.1.

Disposal

A number of methods can be utilized to treat and dispose of the refuse generated in cities. But the feasibility of the method depends on the availability of sources, market condition for refuse generated products, transport and labour availability. More than 90 per cent of solid wastes are disposed of in land filling in our country.

TABLE 18.1: **Solid Wastes Generation in Major Cities**

Town	Population (1981) (million)	Solid wastes		Collection efficiency (%)
		generated (tonnes)	collected (tonnes)	
Bombay	8.23	3200	3100	96.6
Madras	4.28	1819	1637	90.0
Bangalore	0.29	1800	1225	68.1
Ahmedabad	0.25	1200	1080	90.0
Kanpur	0.17	2142	1500	70.0
Pune	0.17	1000	700	70.0
Lucknow	0.10	600	500	83.3
Total		117.61	9742	82.8

Source: Kala and Khan, 1994.

Compositing of vegetable wastes and byproduct of other agricultural industries is very ideally suitable for conversion into organic manure which is very rich in nitrogen, phosphorus and potash. Incineration is a process of controlled combustion for burning solid, liquid and gaseous combustible wastes to gases and residue containing non-combustible material. Refuse Derived Fuel Production is very ideal method to produce an improved solid fuel or pellets from refuse originating from house holds, commerce and industry. This process completes in four steps. First, the shredding of all waste is must. Second, separation of light waste materials. Third, drying of separated waste material and finally compression into pellets of dried materials.

Uses of Oil Extracted from Oilseed Trees

Since, India has a vast potential source of forest trees. About 31 species of forest origin are important oil bearing. The quantities and the method of their utilization have been presented in Table 18.2 and 18.3.

TABLE 18.2: **Total Utilization of Seeds**

S.No.	Tree	Seed production	Oil content (%)	Oil production tonnes
1.	Mahuwa	71,428	35	25,000
2.	Neem	1,00,000	20	20,000
3.	Karanju	25,900	37	7,000
4.	Kusum	9,900	33	3,000
5.	Sal	25,000	12.5	3,125
6.	Kokum	500	40.0	200
7.	Khakan	1,637	46.5	625

TABLE 18.3: **Uses of Oil Extracted from Oilseed Trees**

Seeds and oil content	Botanical name	State	Uses
Tobacco (25%)	*Nicotiana tobaccum*	A.P.	Edible
Mahuwa (35%)	*Bassia latifolia*	M.P., Bihar, Gujarat, M.S., A.P., U.P., W.B., Karnataka	Laundry soap, chocolate, cooking, grease, confectionery purposes and stearic acid
Sal	*Shorea robusta*	M.P., U.P., Bihar, Assam W.B.	Cooking, lighting cocoa, butters, substitute
Neem (12-13%)	*A. indica*	U.P., Bihar T.N., M.S., Gujarat Delhi and Karnataka	Toilet and laundry soap medicines uses and pesticidal uses

Recycling of Sugarcane Press Mud

Sugar industry is one of the most leading industries of India. About 400 small and large surface factors are in working order. They precipitate otherwise down at the bottom of clarifier remained after processing cane sugar is known as pressmud. Although, pressmud is used efficiently as manure and biogas feed stock in rural areas but due to high sucrose content (5-15% sucrose on dry weight basis), it can be used as an inexpensive substrate in fermentation process for production expensive use in a variety of industries and 35000-40000 tonnes is currently manufactured per annum (Bai and Malligai, 1995).

Bio-Conversion of Maize and Sorghum Crops into Alcohol and Biogas

Energy is the basic element of modern technological break through for progressive socio-economic development. It plays a very vital role in all conceivable areas like domestic, agricultural, industrial and transport sectors. Efforts are being made world over by the scientist for suitable alternate energy sources. Out of various energy sources like solar, wind, tidal and energy from biomass, the later is by far the most promising because it is convenient, economical and essentially inexhaustive. Thus a long-term practical solution to meet energy requirement is in the conversion of non-fossil carbon i.e. biomass into fuel. The photo-energy converted as bio-energy in the form of ethanol and biogas can be utilized efficiently as a fuel. The biomass of sorghum and maize is identified and used for increasing the ethanol and biogas production in country.

Ethanol is one of the best liquid fuels as it is relatively superior to gasoline (petrol), safer to handle, burns cleaner, causing less pollution and has excellent cetane number (around 106) as evidenced by the Brazilian National Alcohol Programme, 1975 and Gasohol programme in USA, 1978. An amount of 1044.5 million US dollars can be saved by using or mixing 20% ethanol in petroleum products, as expenditure incurred for importing petroleum products of 5222.7 million dollars from OPEC countries during 1993-94. Maximum ethanol yield 2455 l/ha was produced by 100000 plants/ha in case of maize while it was higher (2629 l.ha) in case of sweet sorghum with plant density 120000 plants/ha.

Biogas has been identified as another potential energy source especially for rural area. It can be used for cooking, lighting, running engines or generation of electricity. Maximum gas yield 267.43 l/kg d.m. was recorded after 63 days in case of fresh stalk of sweet sorghum which was much higher than that obtained in case of fresh cow dung (223.64 l/kg d.m.).

Ethanol Potential of Maize and Sorghum Crop

At Agro energy unit of IARI, the highest ethanol yield (96,406 and 1888 l/ha, respectively) by plant population @ one lakh/ha in rabi 1989-90 when juice, stalk and grain were fermented while it was maximum (102, 416 and 2195 l/ha, respectively) when nitrogen was applied @ 150 kg N/ha (Table 18.4).

The ethanol yield from sweet sorghum juice, stalks and grains was in increased significantly by increasing plant population. The higher ethanol yields in summer 1990 were 1146, 1815 and 814 from juice, stalk and grain, respectively with the plant population 120,000 plants l/ha. Nitrogen application also increased significantly the ethanol yield from all feedstocks. Maximum ethanol was obtained with N application @ 100 kg N/ha in order 1224, 1841 and 843 from juice, stalk and grain, respectively.

Biomass Production from Maize and Sorghum Stalks

Preliminary digestions were conducted at IARI to evaluate the suitability of fresh and the extracted stalks of sweet sorghum and maize for biogas production. Since, these materials failed to produce any noticeable gas when digested anaerobically. These were then modified by mixing with cowdung and fermented anaerobically for biogas generation. The stalks of maize and sweet sorghum collected before and after the juice extraction were dried and finely powdered to pass a 40-60 mush sieve: These were then mixed with cowdung in the ratio of (1:1) on dry weight basis. The data in the gas yields obtained from cowdung and its mixtures with stalks of sweet sorghum and maize show that the supplementation of cowdung to these stalk (fresh as well as juice extracted) improves the conversion efficiency of the materials in terms of energy recover as biogas (Table 18.5). Highest amount of biogas is obtained from a mixture of cowdung and fresh stalks of sweet sorghum (267.43 l/kg d.m.) followed by maize (255.6 l/kg d.m.). Likewise, the supplementation of spent stalks of sweet sorghum and maize to cowdung gave 244.91 and 186.02 l/kg d.m., respectively. It is observed that the gas yields from all mixtures except the one containing juice extracted maize stalks were higher than the control (cowdung). It is interesting to note that, the sole stalks when digested alone, did not produce any biogas because of the obvious reasons of unfavourable environment

TABLE 18.4: **Biomass and Ethanol Production of Maize and Sorghum as Influenced by Plant Population and Nitrogen Levels**

Treatments	Biomass yield (t/ha)		Ethanol production (litre/ha)							
	Maize	S.Sorghum	Maize				Sweet Sorghum			
	(Rabi)	(Summer)	Rabi 1989-90				Summer 1990			
	1989-90	1990	Juice (1)	Stalk (2)	Grain (3)	Total (2+3)	Juice (4)	Stalk (5)	Grain (6)	Total (5+6)
Plant population (plant/ha)										
80,000	10.16	12.76	78	302	1835	2137	780	1429	760	2099
100,000	11.99	15.44	81	376	2079	2455	1166	1748	724	2472
120,000	11.76	16.31	96	406	1888	2294	1146	1815	814	2629
CD at 5%	0.840	1.493	11.5	22.9	191.0	200.0	80.0	109	NS	220
Nitrogen levels (Kg/ha)										
0	8.83	11.97	60	270	1485	1755	743	1395	517	1912
50	11.84	14.99	81	376	1879	2273	1138	1705	731	2436
100	12.64	16.46	93	404	2158	2562	1224	1841	843	2684
150	13.22	15.92	102	416	2195	2611	1114	1716	853	2569
CD at 5%	1.729	0.29	6.5	25.4	77.3	108	76	250	110	199

Source: **Saheb**, 1991.

for the micro-organisms to use the materials as a source of carbon and energy.

Composition of Biogas

The composition of biogas in the initial stages of digestion comprise mainly of carbon dioxide and residual nitrogenous gases but with the progress of digestion the proportion of methane steadily increase. Because in the initial stages of digestion, a group of facultative micro-organisms act upon organic substrate and convert into soluble monomers which are then converted to organic acids.

These acids primarily acetic acid are substrate for the second group of anaerobic bacteria (methogenic) which convert them eventually to methane and carbon dioxide (Table 18.6). At the steady stage of digestion, the optimum concentration of methane and carbon dioxide in biogas from the different treatments varies between 55-56 percent and 34-33 per cent, respectively. The biogas sample from a mixture of cowdung and the fresh stalks of sweet sorghum contained as high as 65% methane and 32% carbon dioxide.

TABLE 18.5: **Gas Yields from Cowdung and Mixture of Cowdung with Freash and Extracted Stalks of Sweet Sorghum and Maize**

	Gas yield litre/kg dry matter				
Period (days)	CD	CD+ SSI	CD+ SSII	CD+ MI	CD+ MII
42	195.84	227.28	187.11	186.41	136.42
49	206.91	241.78	208.41	224.16	160.32
56	216.31	254.28	226.81	241.51	174.12
63	223.44	267.43	244.91	255.16	186.02

Source: Saheb, 1991.

Even after juice extraction, sweet sorghum produced more biomass compared to cowdung. It means, the juice can be used for ethanol production while the bagasse can be used for biogas production which renders efficient utilization of the biogas.

TABLE 18.6: **Biogas Composition of Cowdung Mixture with Fresh and Extracted Stalks of Sweet Sorghum and Maize**

Treatments	Methane (%) (V/V)	Carbondioxide (%) (V/V)	Nitrogen and other gases (%) (V/V)
Cowdung	55	39	6
CD+SSI	65	32	3
CD+SSII	57	36	7
CD+MI	59	34	7
CD+MII	57	36	7

CD — Cowdung
SSI — Fresh sweet sorghum stalks
SSII — Juice extracted sweet sorghum stalks
MI — Fresh maize stalks
MII — Juice extracted maize stalks

Source: Saheb, 1991.

Suggestions in Efficient Utilization of Agro-Industrial by Products and Wastes

1. The dung and crop wastes are the major domestic fuel providing as much as 90% of house hold energy in many villages and a considerable proportion in urban areas of our country. Therefore, agroforestry/farm forestry system/social forestry should be adopted so that cowdung may be used for composing and recycling back to soil fertility.
2. Provision of socio-economic and institutional education should be open on large scale so that proper collection of wastes may be possible.
3. Installation of gobar gas plants in each village on large scale with proper guidelines to farmers.
4. Proper collection of liquid and solid wastes and improve the compost plants in the urban areas.
5. Research and experimental work should be aimed to accelerate the mineralization of wastes in the soil to meet the crops at the required time.
6. Resources survey at national level.
7. Technology planting for sustainable use.
8. Establishment of an information and documentation centre.

Conclusion

It is a debatable question whether by products and wastes should be called as waste material or some other suitable term should be replaced. Strictly speaking, there is no material in the world which can be called as waste. Of course, we will have to exploit them by using latest scientific and technological innovations, so that residues can become more valuable as the main products.

REFERENCES

Kala, J.C. and Khan, R.R. (1994). Management of municipal solid wastes. *Yojana* 37 (24): 18-20.

Kasturibai, R. and Malligai, T.S. (1995). Recycling sugarcane pressmud for the production of lactic acid through fermentation of lactobacillus SP. *Proceeding of* "National Conference on Conversion and Utilization of Biomass" at GHREPD, on November 16-17, 1995, Bakoli, New Delhi: 58-64.

Saheb, S.D. (1991). Effect of plant populations and nitrogen levels on biomass yield of maize and sweet sorghum and its utility as ethanol and biogas production, Ph.D. Thesis, IARI, New Delhi.

Singhal, O.P. and Solanki, R.C. (1993). Power production through gasification of farm rasidues. *Proceeding of* "National Conference on Conversion and Utilization of Biomass" at GHREPD, on November 16-17, 1995, Bakoli, New Delhi: 82-90.

19

A CASE FOR BIOGAS BASED RURAL INDUSTRIALISATION

—V.K. VIJAY, RAJENDRA PRASAD AND JAGPAL SINGH

Rural Industrial Complex

The proposed rural industrial complex will be a unique example of local resource utilization, decentralised energy generation, diversified activities, environmental friendliness employment generation and self sustenance. It takes care of backward and forward linkages of production activities. The activities of the complex will be centred around the biogas plant and the dairy unit. Biogas provides them the energy required and dairy feeding them with the raw materials.

Large size biogas production require large amount of dung which poses dung collection problems in rural areas. This problem is solved by installing dairy in the complex. Dairy will also supply raw materials for milk processing industry and leather processing unit.

Food processing and other industries can be taken up in the complex considering availability of raw materials and market opportunities. Such a complex can be set up in a cluster of villages having population around 20-25 thousand. The complex can be developed on waste land available near the village.

Dairy Unit

Based on milk demand in the area and cowdung requirement for biogas plant, a dairy unit can be planned consisting of 250 cattle. It is assumed that 200 cattle will be milching at any point of time. The total investment for such a dairy unit will be around 30 lakh including cost of cattle, cattle shed and storage etc. The net profit worked out is about Rs. 1,25,000 per month. The dairy will generate employment for 45 people.

Biogas

Biogas is directly linked with dairy unit for the dung supply. The gas generated is supplied to various industrial production units including the dairy itself. Waste affluent from these production units can also be circulated back as a feed material to biogas plant. Biogas manure can be used for agriculture and waste land development which in turn will provide the fodder for the cattle. The biogas can also be linked to toilets if possible and thus helps in sanitation also.

Considering energy requirements of the production units, biogas plant of 145 Cu m capacity can be installed for the purpose. This will required 3625 kg of dung every day which can easily be supplied by the dairy unit. The biogas system will provide employment to 4 persons. Investment cost to set up the system is about Rs. 5.0 lakh. Net profit generated from biogas unit is estimated to Rs. 16,000 per month.

Flour Mill

Flour is an essential commodity of food in most of the households in India. Flour mills are run using electricity or a diesel engine. Erratic and low voltage supply of electricity creates problems. In diesel operated flour mill, cost of milling is high. Therefore, biogas operated dual fuel engine based flour milling can be more reliable and economical.

A 10 hp flour mill may be sufficient for this activity for local needs. The milling capacity of the mill is about 125 Kg/hr. The product of this unit can easily be marketed at local level. 40m^3 biogas will be required to run the activity. The cost of this unit works out at 2 lakh rupees with a net profit Rs. 10,000 p.m. It can provide employment to 4 persons.

Fruits and Vegetable Processing Industry

Fruits and vegetable processing unit can be a part of the rural industrial complex. Selection of Fruits and vegetables to be

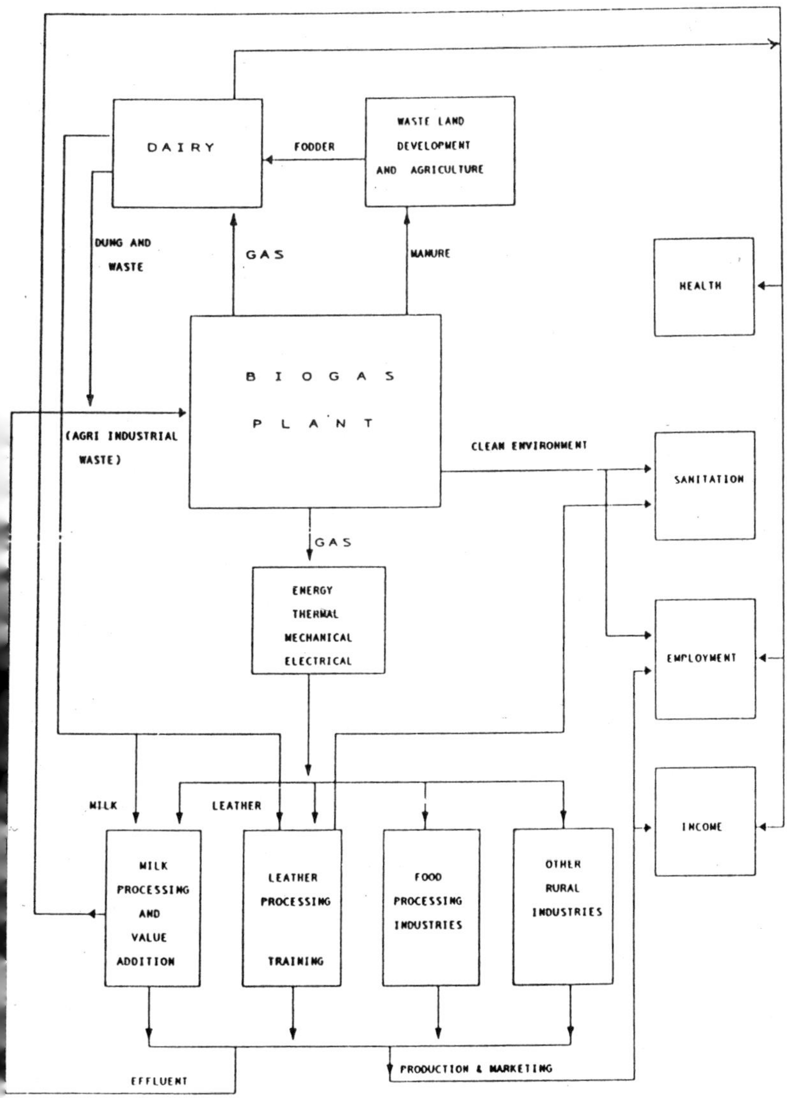

Fig. 19.1: Linkages of Dairy Supported Large Size Biogas Plant for Rural Industrialisation

DETAIL OF RURAL INDUSTRIAL COMPLEX

S. No.	Unit/Enterprise	Employment Generation in Nos.	Energy Supply/Re-quirements	Total Investment (Rs.)	Monthly Profit (Rs.)
1.	Dairy	45	250 cattle	30 lakhs	125,000
2.	Biogas	4	145 cum/day Biogas generation	5 lakhs	16,000
3.	Flour & Cereals milling	4	10 HP (40 cum Biogas daily)	2 lakhs	10,000
4.	Fruit & Vegetable processing	5	30 cum biogas daily	2 lakhs	12,000
5.	Milk processing	3	20 cum biogas daily	1 lakh	4,000

processed will depend on their availability in the area. Tomato being a versatile vegetable, available in many parts of the country can be taken here as an enterprise. Product may also vary depending on the seasons and availability of raw material. Other activities like pickle, jam, jelly etc., may also be taken up.

The surplus production of tomato during seasons could be suitably processed. The total investment cost of the fruit and vegetable processing unit is worked out as Rs. 2.0 lakh with the monthly net profit of Rs. 12,000. The unit will provide employment to 5 persons.

PART EIGHT

BIO-FERTILISATION

20

BIO-FERTILIZERS
The GSFC Approach

—S.B. KUTE AND B.J. PATEL

When the world is approaching very fast towards 21st century, India is neither way far behind. Indian industries are coming up with advanced technologies and at the same time Indian agriculture is also blooming at the same speed. To accelerate the growth of Indian agriculture, the latest research in the field must reach to the remote villages. Agricultural scientists, agricultural universities and industries need to work hand in hand for the desired progress of the rural areas.

Looking to the increasing human population, increase in food grain production is the ultimate goal of the scientists. The thought is required to be given for monitoring the prevailing nutrient supply system. Fertilizer use efficiency also needs to be increased to a considerable extent. The scientists have conceived the thought of Integrated Plant Nutrient Supply System (IPNS) to increase the agricultural productivity. The concept of IPNS is the balanced application of Chemical Fertilizers, Biofertilizers, Organic Manures, Green Manures, Crop residues etc. Chemical fertilizer production is an energy intensive process and utilises the fast depleting non renewable sources of energy. Organic manures are bulky in nature, not available to the required extent and costly too. Considering this, the use of Biofertilizers in judicious combinations with Chemical fertilizers is the ultimate, important and low cost input yet promising alternate.

The term Biofertilizer itself denotes that it is a 'Live Fertilizer'. The typical Nitrogen Fixing and Phosphate Solubilizing attribute of certain soil bacteria is the key phenomenon to use them as

Biofertilizers. The free Nitrogen content (78%) of our atmosphere is no way useful to the crop. Certain soil bacteria have got an enzyme called Nitrogenase which is capable of fixing this nitrogen to the soil and further because of certain chemical and biochemical reactions the fixed nitrogen is converted to such a form which can be absorbed by the plant roots. Certain soil bacteria release organic acids which dissolve the chemically bound unavailable phosphates of the soil and make them available. With these phosphate solubilizing bacteria, optimum utilization of applied phosphatic fertilizer can be achieved.

Using these fascinating attributes of the bacteria, Gujarat State Fertilizers Company Limited (GSFC), Vadodara heartfelt the need of developing different Biofertilizers. Fertilizer Association of India, New Delhi has also boost up GSFC to make a breakthrough amongst all the fertilizer industries by commercializing Biofertilizers to the service of farming community. After considerable experimentations from 1980 to 1984, GSFC successfully launched three types of Nitrogen fixing Biofertilizers in 1984. Phosphate solubilizing Biofertilizer was introduced in 1987 and recently in 1995 a new nitrogen fixing Biofertilizer was introduced specially for sugarcane.

At present GSFC is producing five different types of Biofertilizers for various crops.

1. Sardar Azotobacter Culture : For cereals, cash crops and vegetables.
2. Sardar Azospirillum Culture : For cereals, cash crops and vegetables.
3. Sardar Rhizobium Culture : For pulses and leguminous crops.
4. Sardar Phosphate Culture : For all the crops.
5. Sardar Super Culture : For sugarcane crop.

The commercialization of Sardar Biofertilizers involve following major steps (Chart : I).

(A) Search for efficient strains.
(B) Production and Quality Control.
(C) Promotion of Biofertilizers.
(D) Marketing of Biofertilizers.

CHART-I

BIOFERTILIZER PRODUCTION AND QUALITY CONTROL AT GSFC

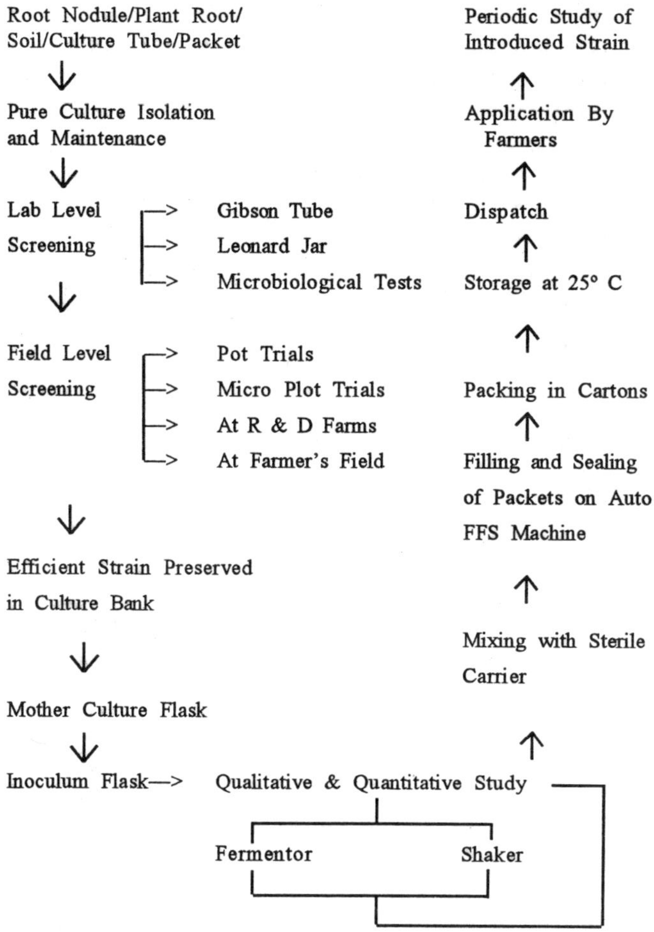

(A) Search for Efficient Strains

Various bacterial strains exhibit various characteristics. It is required to study these characteristics in depth for finding out the efficient strain in pure form. The search is carried out in two stages as under.

(i) Isolation and Identification

Depending upon the growth requirement of specific bacterial strain, synthetic media are prepared and sterilized in the laboratory. Azotobacter, Azospirillum and Phosphate Cultures are isolated on these media from the rhizosphere of the plant root system while Rhizobium Cultures are isolated from the well developed, healthy nodules on the root system of leguminous plants. Soil and root samples are collected from the different regions of Gujarat state for isolation. The cultures are also brought from various research institutions like IARI, ICRISAT and NIFTAL. On the basis of morphological, colonial and biochemical characteristics, the strains are identified and differentiated. The pure cultures are preserved for screening of the efficient strains.

(ii) Screening of the Pure Isolated Strains

The screening of the pure isolated strains is carried out at two different but inter related levels.

(a) Lab level.
(b) Field level.

(a) Lab Level Screening

Azotobacter and Azospirillum cultures are tested in a special device called Gibson's Tubes. In Gibson's tube, the seeds of various cereal crops are grown on nitrogen free media and the culture treatment is also given. Depending upon the nitrogen fixing ability of the culture, the growth of the plant will vary.

Rhizobium cultures are tested in a device called Leonard Jars for nodulation studies. Nitrogen fixation studies of all the these cultures are carried out by Acetylene-Ethylene Reduction Assay.

The Phosphate cultures are studied on Tricalcium Phosphate Agar. When this culture is grown on tricalcium phosphate agar, it liberates certain organic acids which are diffused into the medium. The acids attack on the bound phosphates i.e., tricalcium phosphate and the P O is released from the complex into the medium. The phosphate solubilization is observed in the form of zone of clearance around the colony of phosphobacteria developed on tricalcium phosphate agar.

The strains shown satisfactory results are further tested at field level.

(b) Field level screening

Field level testing of all the cultures is carried out every season in pots, microplots and at our own Research and Demonstration Farm. An experiment conducted on Cotton crop consecutively for three years i.e., 1990 to 1992 (Table 20.1) revealed that, application of only chemical fertilizer have given yield of 1190 Kg./ha. When various Biofertilizers are applied alongwith the chemical fertilizers the yield has been increased. Maximum yield of 1361 Kg./ha has been obtained with the application of recommended dose of chemical fertilizers and Azotobacter Biofertilizer, but when the recommended dose of chemical fertilizers has been applied alongwith Azospirillum

TABLE 20.1: **Effect of various biofertilizers over and above the recommended dose of chemical fertilizers on cotton crop**

(Yield in Kg./ha.)

Treatments	Year			Average of Three Years
	1990	1991	1992	
Fertilizer	1734	512	1323	1190
Ferti. + Azoto.	2034	425	1623	1361
Ferti. + Azosp.	1896	546	1419	1287
Ferti. + Phos. Cult.	1810	550	1469	1276
Ferti. + Azoto. + Phos. Cult.	2161	552	1458	1390
Ferti. + Azosp. + Phos. Cult.	1995	565	1699	1419
Year Mean	1938	525	1498	-
Test	N.S.	N.S.	N.S.	N.S.
SEM	99.26	59.24	160.01	65.260
CD	-	-	-	-
CV %	8.87	19.55	18.50	14.95

Var. Hy-6
Ferti. Dose : 320 N, 160 P/ha.
Phos. Cult. : Phosphate Culture
Azoto. : Azotobacter
Azosp. : Azospirillum

and Phosphate Culture; the highest yield of 1419 Kg./ha was obtained. Similarly in another experiment (Table 20.2), in case of wheat, the yield of 1727 Kg./ha was obtained with the application of Phosphate Culture and chemical fertilizers as compared to 1542 Kg./ha when only chemical fertilizers were applied. The second

TABLE 20.2: **Effect of various biofertilizers over & above the recommended dose of chemical fertilizers on Wheat & Bajra crops**

(Yield in Kg./ha.)

Treatments	Wheat	Bajra	
		Grain	Fodder
C0 N0 PF	552.00	1709.00	8019.00
C0 NF P0	1637.00	1908.00	11562.00
C0 NF PF	1542.00	1783.00	12303.00
C1 N0 PF	770.00	1673.00	10081.00
C1 NF P0	1577.00	2454.00	12077.00
C1 NF PF	1992.00	2108.00	15008.00
C2 N0 PF	690.00	1873.00	10306.00
C2 NF P0	1632.00	2084.00	11369.00
C2 NF PF	2227.00	1818.00	14525.00
C3 NF P0	1723.00	2296.00	10725.00
C3 NF PF	1727.00	2029.00	13494.00
Test.	Signi.	Signi.	Signi.
SEM	196.56	148.14	1086.58
CD	579.86	437.02	3205.42
CV %	23.31	12.99	15.99

	Wheat	Bajra
Var.	Kalyan Sona	MH-179
Ferti. Dose	100 N, 56 P/ha.	75 N, 37.5 P/ha.
Year	1992-93	1992

C0 :No Culture
C1 : Azotobacter
C2 : Azospirillum
C3 : Phosphate Culture
N0 : No Nitrogen
NF : Full Nitrogen
P0 : No Phosphate
PF : Full Phosphate

highest yield of 1992 Kg./ha was obtained in a treatment where Azotobacter Biofertilizer was applied alongwith the chemical fertilizers. The highest yield of 2227 Kg./ha was obtained with the application of Azospirillum and chemical fertilizers. In case of Bajra the highest yield of 2454 Kg./ha has been obtained when Azotobacter Biofertilizer was applied alongwith the Nitrogenous fertilizers, while the second highest yield of 2296 Kg./ha was obtained with the application of Phosphate Culture alongwith the Nitrogenous fertilizers. However, in respect of fodder production, the significant increase due to the application of various Biofertilizers were not obtained over and above the application of chemical fertilizers. They have given significantly higher yield over that of only phosphorus application through chemical fertilizers. The experiment carried out on Mustard crop for two years i.e., 1991-92 and 1992-93 (Table 20.3) revealed that, the highest yield of 1131 Kg./ha was obtained with chemical fertilizers and Azotobacter Culture as compared to 884 Kg./ha when chemical fertilizer alone was applied. The second highest yield of 1033 Kg./ha was obtained with application of Azospirillum over and above the chemical fertilizers.

TABLE 20.3: **Effect of various biofertilizers over and above the recommended dose of fertilizers on mustard crop**

(Yield in Kg./ha.)

Treatment	Year		Average of Two Years
	1991-92	1992-93	
C0 N0 PF	1027.00	268.00	648.00
C0 NF P0	1074.00	388.00	731.00
C0 NF PF	1196.00	571.00	884.00
C1 N0 PF	1037.00	381.00	709.00
C1 NF P0	1285.00	474.00	879.00
C1 NF PF	1528.00	735.00	1131.00
C2 N0 PF	1026.00	460.00	743.00
C2 NF P0	1140.00	498.00	819.00
C2 NF PF	1406.00	659.00	1033.00
C3 NF P0	1182.00	486.00	834.00
C3 NF PF	1359.00	587.00	973.00
Year Mean	1206.00	501.00	-
Test.	Signi.	Signi.	Signi.
SEM	109.15	54.24	57.23
CD	322.00	160.00	162.66
CV %	15.68	18.77	17.50

Var. T-59
Fertilizers 40 N, 40 P/ha.

C0 : No Culture
C1 : Azotobacter
C2 : Azospirillum
C3 : Phosphate Culture

N0 : No Nitrogen
P0 : No Phosphate
NF : Full Nitrogen
PF : Full Phosphate

In microplot trials the cultures are tested in the soils brought from different regions of Gujarat. Trials are also taken on farmer's fields, Government seed farms, Agriculture Universities and various Research Institutions. The trials conducted at various sugar factories in Gujarat (Table 20.4) indicate increase in number of tillers, brix, pol reading and purity percentage with the use of Azotobacter Culture and Phosphate Culture, the phosphate content is also increased from 221 mg/lit to 250 mg/lit. With the increase in phosphate, crop production is increased and at the same time sugar recovery is also increased. In another experiment on Groundnut (Table 20.5) conducted at Chikhli, Dist. Bulsar; an appreciable increase (From 24.8 Kg./Acre to 62.8 Kg./Acre) in available phosphorus was recorded with the treatment of Phosphate Culture.

TABLE 20.4: **Field trials of Sardar Biofertilizers taken alongwith the recommended dose of chemical fertilizers on sugarcane at different locations of Gujarat**

Sr. No.	Parameters	Control RD	RD + AB	Control RD	RD + PC
1.	No. of Tillers	22	25	24	27
2.	Brix	20.89	21.96	21.56	22.49
3.	Pol	17.98	19.86	19.20	20.40
4.	Purity %	86.07	90.44	89.06	90.70
5.	P_2O_5 mg/lit	-	-	221.12	249.48

RD : Recommended dose of Chemical Fertilizers
AB : Sardar Azotobacter Culture
PC : Sardar Phosphate Culture

The parameters studied in all the experiments are germination percentage, green weight and crop yield etc. The strains (from each culture) which are proved efficient after series of tests at lab level

TABLE 20.5 : **Results of the soil analysis to study the effect of Phosphate Culture**

Crop : Groundnut Location :Chikhli (Dist. Bulsar)

Time of Sampling	Organic Carbon (%)	P_2O_5 Kg./acre	K_2O Kg./acre	pH
Before Sowing	0.60	24.8	150.0	8.4
After One month	0.56	24.8	140.0	8.2
After Two months	1.10	68.2	110.0	8.2
After Three months	0.71	58.4	90.0	8.0
After harvesting the crop	0.78	62.8	90.0	7.8

and field level are selected for production.

(B) Production and Quality Control

The production of the selected strain is carried out in following sequence.

Preserved Culture on agar slants —> Mother Culture Flask —> Stock Culture Flask —> Bulk Fermentation on shakers —> Mixing with sterile carrier —> Filling, Sealing and Packing.

(a) *Agar Slants*: The strain proved most efficient throughout all the above tests are carefully preserved in pure form in the refrigerator on agar slants for the use as starter source.

(b) *Mother Culture Flask*: A pin drop of bacterial growth is transferred (under sterile condition) from the agar slant to pre-sterilized liquid medium in flask (Capacity : 750 ml.) for multiplication. The flask is kept on shaker to provide aeration and agitation required for bacterial growth. This flask is called 'Mother Culture Flask' and it is prepared to avoid frequent disturbance to the agar slant preserved above. After the desired growth it is checked qualitatively and quantitatively on microscope and then preserved in refrigerator.

(c) *Stock Culture Flasks*: A small volume of fermented broth is transferred (under sterile condition) to number of small flasks (Capacity : 500 ml.) depending upon the requirement. The mother culture flask is checked for purity on microscope before the transfer. The small flasks are kept on shaker as above for growth. These flasks are called 'Stock Culture Flasks' or 'Inoculum Flasks'.

(d) *Bulk Fermentation on Shakers*: Bulk fermentation is carried out in big flasks (Capacity : 5.0 ltrs.). Medium is prepared and sterilized according to the requirement. The stock culture flasks are checked first qualitatively and quantitatively on microscope. After the confirmation of purity a small volume of fermented broth is transferred (under sterile condition) to big flasks. The big flasks are kept on shaker for the desire growth of the bacteria.

Strict quality control measures are observed right from the agar slants to bulk fermentation on shakers. Periodically calibration of the sterilizers is also carried out with the use of bacteriological indicators. Fermented broth is strictly checked quantitatively for count (i.e. number of bacterial cells/ml. of broth) and qualitatively for purity by staining the bacterial cells before further processing.

(e) *Mixing with Sterile Carrier*: After the fermentation for required hours, when the optimum count of pure desired bacteria is achieved; the broth is mixed with sterile carrier i.e., Lignite and Calcium Carbonate (Calcium Carbonate is added for pH neutralization of lignite). It is mixed in a specific proportion to maintain the proper moisture content i.e., 25% to 30% and optimum bacterial count i.e., minimum 10 per gram of the finished product. This mixing is carried out in a specialized mixing machine. All the necessary precautions are taken to reduce the chances of contamination.

The reasons for using lignite as a carrier in Biofertilizer production.

1. Lignite is cheaper as compared to other carrier material.
2. Easily available from Gujarat Mineral Development Corporation through various agencies.
3. It has got higher water holding capacity.
4. It provides good adhesion to the seeds.
5. It is inert by nature and hence contamination interference is reduced.

(f) *Filling, Sealing and Packing* : To avoid any possibility of aerial contamination, filling and sealing of the culture mixed carrier material is carried out on specially designed, advanced Automatic Form-Fill-Seal Machine. Weighing of the material, bag forming, filling and sealing operations are performed automatically on the same machine at a time with utmost precision. The maximum capacity of this machine is 15-17 bags per minute. The bags are pre-printed with all the necessary information like name of the culture, name of the crop for which it is to be used, the method of application, the dose of application and precautions required to be taken. Manufacturing Date, Expiry Date, Batch No. and Price are also printed on each bag. These bags are further packed in corrugated boxes.

Sardar Biofertilizers are available in following packing sizes.

Name of the Biofertilizer	Available Packing		No. of bags in each corrugated box
1. Sardar Azotobacter Culture	250 Grams	...	50
	1.0 Kg.	...	25
2. Sardar Azospirillum Culture	250 Grams	...	50
	1.0 Kg.	...	25
3. Sardar Phosphate Culture	250 Grams	...	50
	1.0 Kg.	...	25
4. Sardar Rhizobium Culture	250 Grams	...	50
5. Sardar Super Culture	1.0 Kg.	...	25

All the boxes are stored in temperature controlled room.

Finished product is checked periodically for viability and purity of the bacterial cells as well as confirmation of shelf-life.

Production Constraints

(a) *Bacterial Strain:* The efficiency of the bacterial strain in use, depends on various factors like humidity and temperature of the atmosphere, soil pH, soil temperature, chemical and moisture content of the soil and native micro flora of the soil with which the introduced culture has to struggle a lot for the survival.

(b) *Economic viability:* To have a quality product, the instruments and machineries with advanced technology are required which increases the production cost.

(c) *Production process:* Even after utmost care exercised for contamination control, limitation comes at the time of mixing of fermented broth with sterile carrier and at the time of filling, sealing and packing of the bags manually.

(d) *Shelf life:* Because of the limited shelf life of the product only limited production can be taken well in advance.

(C) Promotion of Biofertilizers

The actual consumption of biofertilizers in Gujarat and India is to the extent of 600 and 3000 MTs as against the total potential of about 8000 and 5 lakh MTs respectively. It necessitates the aggressive efforts in respect of promotion of various biofertilizers. GSFC has been promoting its Sardar Biofertilizers scientifically as under and thereby increasing the sale of our biofertilizers from 8500 packets in the year 1983-84 to 16 lakh packets during 1994-95.

(i) Field Experiments on our R&D Farm

Being the new product, there is a need to conduct scientific experiments systematically on various crops in order to transfer this latest technology from lab to land for the benefit of farmers and accordingly GSFC has been conducting such experiments on its own R&D Farm since last 10 years or so. The crop production was increased substantially due to the application of Rhizobium to leguminous crops, Azotobacter to cereal crops, Azospirillum to transplanted crops of paddy and vegetable and Super culture to sugarcane crop over and above the application of recommended doses of fertilizers. The further increase in crop production was noted when Phosphate culture was applied.

(ii) Trials on Talukas and Sugar Factory Farm

Large number of trials were undertaken on various crops at Government Taluka Farms. The crop production was increased in respect of almost all the crops due to application of Azotobacter, Azosprillum and Phosphate Culture. The highest yield increase was achieved in case of Sugarcane (20 to 21%) due to application of Azosprillum and Azotobacter which was followed by Groundnut (18%) due to the application of Phosphate Culture. The highest straw production was achieved in case of wheat (12%) due to application of Azotobacter (Table 20.6)

TABLE 20.6: **Trials taken at Government Farms (1986-87 & 1990)**

(Yield in Kg./ha.)

Crop	Year	Locations	BF	RD		RD + BF		% Increase	
				Grain	Straw	Grain	Straw	Grain	Straw
Paddy	86-87	2	AB	4000	2000	4320	2100	8	5
Bajra	86-87	9	AB	1894	12372	1967	12888	4	5
			ASP	1894	12372	2000	12942	6	5
Wheat	86-87	7	AB	2374	1432	2615	1587	11	12
			ASP	2374	1432	2578	1494	9	5
Must-ard	86-87	6	AB	910	-	956	-	5	-
			ASP	910	-	1025	-	13	-
Sugar-cane	86-87	9	AB	88700	-	97360	-	21	-
			ASP	88700	-	96710	-	20	-
Grou-ndnut	90	4	PC	785	-	930	-	18	-

AB : Azotobacter

ASP : Azozpirillum

PC : Phosphate Culture

RD : Recommended dose of Chemical Fertilizers

BF : Biofertilizers

In the adoptive trials conducted in collaboration with "Vasantdada Sugar Institute, Pune (Maharashtra)" in different agro climatic conditions, substantial increase in germination, tillering and yield was observed due to the application of various biofertilizers. The cane yield was increased from 82 MTs to 100 MTs, 91 MTs and 90 MTs per hectare due to the application of Phosphate Culture, Azotobacter, Azospirillum respectively over and above the application of recommended dose of chemical fertilizers indicating the beneficial effect of biofertilizers (Table 20.7).

TABLE 20.7: **Results of the adoptive trials conducted on Sugarcane in collaboration with Vasantdada Sugar Institute, Pune; During the year 1991-92 and 1992-93 in different Agroclimatic conditions. (Average of Five Locations)**

Sr. No.	Parameters	Control RD	RD + AB	RD + ASP	RD + PC
1.	Germination %	68.35	71.92	71.50	73.12
2.	Tillering Ratio	2.06	2.23	2.39	2.04
3.	Cane Yield (MT/ha.)	81.60	90.77	89.88	100.10

RD : Recommended dose of Chemical Fertilizers
AB : Sardar Azotobacter Culture (@ 5.0 Kg./ha.)
ASP : Sardar Azospirillum Culture (@ 5.0 Kg./ha.)
PC : Sardar Phosphate Culture (@ 5.0 Kg./ha.)

(iii) Demonstration on Farmer's Fields

Numerous demonstrations conducted with various crops on farmer's fields on biofertilizers, convincingly showed the increase in crop production. The highest increase was observed in case of Maize (38%) due to the application of Azosprillum and Phosphate culture followed by 30%, 28%, 23%, 22%, and 21% in case of bajra and gram, paddy, potato and wheat, mustard and castor by the application of AB+PC in bajra, Rhi+PC in gram, ASP+PC in paddy, AB in potatoes and AB+PC in wheat, ASP+PC in mustard and AB+PC in castor respectively. The per hectare dose of biofertilizer was around 750 grams (3 packets) in case of all the crops except sugarcane where the dose was around 5 Kg/hectare (Table 20.8).

TABLE 20.8: **Demonstrations taken on farmers field in Gujarat State**

Average Yield in Kg./ha.

(Years : 1990-91, 1991-92, 1992-93 & 1993-94).

Sr. No.	Crop	Loca-tions	BF	Grain RD	Grain RD+BF	% increase
1.	Wheat	48	AB	3774	4246	13
2.	Wheat	33	AB+PC	3013	3690	23
3.	Wheat	49	ASP	3923	4543	16
4.	Wheat	28	ASP+PC	3105	3738	20
5.	Bajra	2	AB+PC	2363	3065	30
6.	Bajra	9	ASP	2498	2880	15
7.	Bajra	6	ASP+PC	2078	2638	27
8.	Paddy	17	ASP	3836	4565	19
9.	Paddy	16	ASP+PC	4331	5520	28
10.	Maize	3	ASP+PC	2513	3475	38
11.	Cotton	3	AB	3750	4375	17
12.	Cotton	11	AB+PC	2148	2543	18
13.	Mustard	9	ASP	1495	1800	20
14.	Mustard	9	ASP+PC	1385	1685	22
15.	Mustard	4	AB+PC	1643	1995	21
16.	Castor	8	AB+PC	2638	3185	21
17.	Sugarcane	4	ASP+PC	175000	210000	20
18.	Potatoes	3	AB	12793	15708	23
19.	Chillies	4	AB	7625	8625	13
20.	Gram	3	Rhi	1000	1275	28
21.	Gram	4	Rhi+PC	953	1238	30
22.	Groundnut	19	Rhi	1429	1622	14
23.	Groundnut	22	Rhi+PC	1293	1538	19

AB : Azotobacter
ASP : Azospirillum
PC : Phosphate Culture
RHI : Rhizobium

R D : Recommended dose of Chemical Fertilisers
B F : Biofertilizers

In addition to the above, number of following promotional

activities are undertaken—

(a) Demonstration meetings on farmer's fields,
(b) Seasonal campaigns,
(c) Farm Youth Training Programmes,
(d) Audio Visual Display for educating farmers,
(e) Literature on Biofertilizers in local languages
(f) Distribution of paper stickers to our FIC-cum-Depots, Institutional agencies, dealers etc.
(g) Publicity through local newspapers and our Gujarati Krishi Jivan magazine which has circulation over 75000 copies amongst farmers.
(h) Publicity through exhibitions, fairs, festivals etc.

(D) Marketing of Biofertilizers

GSFC's dynamic marketing philosophy is to market biofertilizers with the support of development and promotional programmes and by offering before and after sales services to farmers.

(i) Farm Information Centres-cum-Depots

GSFC through its over 200 "Farm Information Centres-cum-Depots" situated in rural areas and manned by experienced agricultural graduates provide the technical knowhow, conduct the demonstrations and give before and after sales services to farmers and has been instrumental in marketing Sardar Biofertilizers alongwith other agro products including fertilizers in Gujarat State.

(ii) Distribution

In order to provide our biofertilizers upto the village level, GSPC has appointed its own Distributor's network for marketing various types of biofertilizers.

(iii) Co-operative and Institutional Agencies

Priority is also given to the Co-operative Societies and other Institutional Agencies for distribution of the biofertilizers.

(iv) Subsidy

In order to promote the biofertilizers, Government of Gujarat has been offering subsidy to the extent of 50% to the farmers throughout FIC-cum-Depots and Co-operative Societies. It has been helpful in pushing the biofertilizers in rural areas.

(v) Farmer's Survey

In order to know the farmer's response to Sardar Biofertilizer's use, the survey was carried out in 5 talukas of Vadodara District. In all 80 farmers were contacted (Table 20.9). About 62% farmers used Sardar Biofertilizers as they were convinced about their benefits, around

TABLE 20.9: **Farmer's Survey on Sardar Biofertilizers**

Parameters	Feedback from the Farmers		
	Respondents (in Nos.)	Respondents (in %)	Opinion/ Reaction
(1) The reason of use of Sardar Biofertilizers	37	61.67	As they were convinced of the benefits of the Sardar Biofertilizers.
(2) The price of Sardar Biofertilizers.	28	46.67	The prices are reasonable.
	19	31.67	The prices are low.
	13	21.66	The prices are high.
(3) The availability of Sardar Biofertilizers on time.	52	86.67	Timely available.
(4) The Criteria of purchase of Sardar Biofertilizers.	36	60.00	On the basis of quality.
	18	30.00	On the basis of return.
	06	10.00	On the basis of lower price.
(5) The Continuity of use of Sardar Biofertilizers.	48	80.00	Permanently using Sardar Biofertilizers after adopting the same as a part of input.
(6) •Farmers' satisfaction of Sardar Biofertilizers.	54	90.00	Fully Satisfied with the use of Sardar Biofertilizers.

78% farmers found the prices as lower and reasonable, nearly 60% opined that they had gone for them due to better quality, 80% informed that they would continue to use them, 87% indicated about their timely availability and lastly about 90% were fully satisfied about their use.

Marketing constraints

Biofertilizers, being new to the producers, distributors and also farmer consumers, have the following constraints affecting the consumption adversely :

(i) Pricing Policy and Packing

No definite pricing policy exists at present for biofertilizers. Packing standards are also not fixed. This has resulted in the supply of very poor quality biofertilizers having variations in the prices and the quality as well.

(ii) Lack of Awareness

Inspite of the emphasis given on training, education, knowledge adoption and perception, most of the farmers still do not have inclination for the use of biofertilizers. They yet need extensive education and practical training on various aspects of biofertilizers through aggressive promotional programmes, as well as before and after sales services.

(iii) Inadequate Shelf-life

The biofertilizers presently available, have shelf-life of 3 to 6 months. Farming in most of the areas fully depends on nature and such variations in agro climatic conditions result in the change in purchasing power of farmers affecting the procurement of inputs adversely which leads to the unsold stock of biofertilizers at the distribution centres and sales points with shorter shelf life thereby affecting its overall efficiency.

Suggestions

Keeping in view the thrust areas identified and the strategy worked out for increasing the consumption and the marketing of the biofertilizers, the following suggestions are given :

(i) Strict quality control standards in production, storage, distribution and the use must be maintained.
(ii) Uniform packing and pricing policy be set up and subsidy may be provided on capital as well as revenue expenditure.
(iii) Alongwith manufacturers own marketing channels, Government channels may be set up.
(iv) State and Central Government should have facility for testing of biofertilizers as assurance to farmers.
(v) On the line of "Chemical Fertilizers", the "Biofertilizer Control Order" should be worked out to ensure the quality.

Future Stratgey

In order to increase the marketing of biofertilizers substantially, the following strategy has been decided :

(i) Development of marketing network in the neighbouring states.
(ii) Undertaking of "National Biofertilizer Development Programme" as desired by Government of India.
(iii) Introduction of bulk packing.

Conclusion

Experiments, trials and demonstrations convincingly showed the increase in crop production to the extent of 15 to 20% with the use of biofertilizers and the increase was higher on farmer's fields. The additional increase in the crop production by the use of biofertilizers over and above the use of chemical fertilizers may indicate the improvement in fertilizer use efficiency. Inspite of the fact that biofertilizers is a low cost input in increasing crop production, certain limitations and constraints particularly related to shelf-life, high temperature, transportation, timely distribution in remote areas etc., do restrict rapid growth of biofertilizer market. Besides, uniform pricing policy, standard packing and quality test by Government of India/State Governments will help to develop healthy market in the time to come. The last but not the least, there is a great need to work out the aggressive promotional and marketing strategy for increasing the use of biofertilizers as a low cost input supplementary and complementary to chemical fertilizers for getting higher crop production and profit.

Starting with the journey from 8200 packets (2.05 tons) in 1984, during the last year i.e. 1994-95 we have produced 16.20 lakhs packets (i.e. 404.87 tons) (Table 20.10). This eye catching progress from Beneath to Zenith could never have been possible without quality awareness and persperating efforts.

TABLE 20.10: **Quantum of Sale of Sardar Biofertilizers**

(Unit : Tons)

Year	Azoto.	Azosp.	Rhizo.	Phos.	Sup.	Total
1984	00.65	00.37	01.03	-	-	02.05
1985	01.12	00.70	01.75	-	-	03.57
1986	09.09	10.70	12.26	-	-	32.05
1987	13.06	14.84	05.35	-	-	33.25
1988-89	22.62	50.09	32.29	07.50	-	112.50
(15 Months)						
1989-90	21.06	51.21	28.53	21.55	-	122.35
1990-91	21.63	50.35	42.89	19.97	-	134.84
1991-92	43.11	56.52	64.19	55.94	-	219.76
1992-93	55.40	56.70	58.60	95.50	-	266.20
1993-94	67.13	56.33	52.41	104.51	-	280.40
1994-95	72.01	71.07	61.06	174.92	25.81	404.87

Azoto. : Azotobacter Culture
Azosp. : Azospirillum Culture
Rhizo. : Rhizobium Culture
Phos. : Phosphate Culture
Sup. : Super Culture

Tatal Target
1995-96 450 Tons
1996-97 525 Tons

The quality and quantum has rewarded GSFC, with the high

honour of 'The Best Performing Unit in The Country in Biofertilizer Production'; consecutively for two years i.e. 1991-92 and 1992-93, for the year 1993-94 GSFC has been honoured with the 'Second Best Award'—an award from National Productivity Council, Government of India, New Delhi.

21

EXTENSION STRATEGY FOR PROMOTION OF BIO-FERTILIZER USE IFFCO Experience

—G.C. SHROTRIYA

The concept of biological nitrogen fixation through symbiotic bacteria present in root nodules of leguminous plants and trees was known to Indian farmers since centuries. Incorporation of grain and fodder legumes in crop rotations and in agro forestry system was the result of such knowledge. Inoculation of seeds with bacterial culture as an agronomic practice was, however, introduced in a systematic way during mid sixties only. During nineties the concept gained momentum in the wake of threat from depleting natural resources responsible for production of fertiliser nitrogen. In the recent past, for the balanced and efficient plant nutrient management for sustaining the crop productivity and food production IPNS has been introduced. IPNS involves complementary and supplementary use of available plant nutrients through various inorganic and organic sources.

During 1994-95, in all 13.5 million tonnes of major nutrients (N 9.5, P_2O_5 2.9, K_2O 1.1) were consumed for crop production through chemical fertilisers in India. The fertiliser production in the country was only 84% of N and 87% of P_2O_5 consumption. To meet the deficit in consumption and production, about 2.9 million tonnes of urea (46%N) and 0.8 million tonnes of dia-ammonium phosphate (18% N, 46% P_2O_5) was imported. This level of import is likely to continue till we build up capacity of production. The entire K_2O requirement is also imported as the country do not have its own K_2O source. Even with these impressive figures of production and consumption of fertilisers, the fertiliser use in India remains very low at 74 Kg (N, P_2O_5 & K_2O)/ha and it is estimated that about 10 million tonnes of nutrient deficit is created in our soil every year, which is an alarming signal for sustainable agriculture.

Integrated Plant Nutrient Supply System, besides advocating use of alternate nutrient source, also advocate shift in cropping pattern which can change the nutrient depletion pattern in soil. This gives an scope for inclusion of legume in more popular cereal dominated crop rotations. Such a change envisage fixation of nitrogen from atmosphere. Rhizobium inoculation can provide boost to this phenomenon. With the advent of other N fixers like Azotobactor, Azospirillum and azolla etc., N fixation in non-leguminous crops has also become possible. Similarly phospho-bacterium is now available to enhance the availability of soil or fertiliser phosphorus by solubilising the native soil P or fertiliser applied P.

IFFCO's Biofertiliser Use Promotion Strategy

Inspite of so much of advancement in the field of microbiology there has been a lack of demand for the biofertilisers from farmers side, barring some exceptional cases like for soyabean crop in the state of Madhya Pradesh. Certain extension programmes have been introduced to promote the use of biofertilisers. Though the use of biofertiliser packets is increasing, the practice has not gained much popularity as is visible in case of fertiliser and other agro-inputs. The regular extension activities like farmers meetings, field days, literature, demonstrations, exhibitions etc., do help in creating awareness about bio-fertilisers. Besides these activities special strategy has to be worked out for biofertiliser promotion. The components of extension strategy adopted by IFFCO for promoting the use of biofertilisers are as under:

1. Field demonstrations/trials
2. Minikit distribution
3. Intensive area approach
4. Integrated plant nutrient system (IPNS)
5. Enhancing production

Field Demonstrations/Trials

IFFCO initiated promotion of biofertilisers with distribution of culture packets and field trials. Acceptability of Rhizobium culture for legumes is now prevailing at almost satisfactory level among farmers, therefore, trials were initiated with non-legume crops viz., wheat, mustard, sunflower, potato and rice during 1993-94. Some of the exotic strains from Russia was also tested in these trials. The Russian strains tested were Bioflav, Bioagrin, Bioenterin, Biomiz supplied by M/s Biotech International Limited, New Delhi. Results from these trials for wheat, mustard and rice crops are summarised in Table 21.1, 21.2 and 21.3. Yield increases upto 21% in wheat, 40.8% in mustard and upto 9% in rice were recorded when biofertilisers were applied over the recommended doses of fertilisers. When the biofertiliser were applied without application of fertilisers the quantum of increased yield by and large was slightly higher. For example in wheat the yield increase due to biofertiliser application on an average was to the extent of 4.21 q/ha under unfertilised condition compared to 3.41 and 3.12 q/ha under fertilised condition. At higher level of production the yield increase due to biofertiliser application was not quite conspicuous. In mustard also yield increase from 0.83 to 2.19 q/ha was recorded due to biofertiliser application. The differences between the local and exotic strains were not appreciable.

Minikit Distribution

Distribution of minikits of various agroinputs like seed, agrochemicals, fertiliser is common practice for field verification and promotion of newly developed technology. Biofertilisers for the respective crops must be essentially included as one of the component in the minikit distribution programme. IFFCO in its adopted village where critical input distribution is undertaken for rapid promotion of crop production technology has included bio-fertiliser as essential component under different farming situations. This practice has helped in creating awareness and in adoption of biofertilisers use.

Intensive Area Approach

Effectiveness of biofertiliser strains is subjected to certain field

TABLE 21.1: **Results from biofertiliser trials on farmer's fields in Wheat Crop**

(grain yields in q/ha) 1993-94

Location	Absolute Control (A)	(A)+R.S. (B)	Recommen-ded fert. dose (C)	(C)+L.S. (D)	(C)+R.S. (E)
Kotamgaon (Nasik)	9.64	11.42	27.68	32.26	33.40
Mattershem (Hisar)	24.61	27.69	29.33	34.87	35.38
Cordet (Allahabad)	-	-	19.85	-	22.40
Khijirpur (Allahabad)	-	-	39.00	-	41.37
Salarpur (Meerut)	29.60	38.40	55.20	56.80	56.00
(Ludhiana)	33.20	37.00	54.04	55.96	56.20
Jaithal (Ujjain)	24.96	28.54	55.40	-	57.58
Mean	24.40	28.61	40.07 (41.56)	- (44.97)	43.19

R.S. = Russian Strain, L.S. = Local Strain

TABLE 21.2: **Results of biofertiliser trials on farmers field in Mustard crop**

(grain yield in q/ha) 1993-94

Location	Absolute Control (A)	(A)+R.S. (B)	Recommended fert. dose (C)	(C)+L.S. (D)	(C)+R.S. (E)
Hisar	18.46	-	21.54	24.61	26.66
Meerut	7.20	8.68	15.40	15.16	15.20
Burdwan	3.80	4.10	12.00	12.30	12.80
Gwalior	10.04	14.46	17.94	-	23.56
Ludhiana	9.16	10.34	15.04	15.48	15.40
Allahabad	8.18	8.92	8.84	9.40	10.28
Mean	9.47 (7.67)	- (9.30)	15.13 (14.56)	- (15.39)	17.32

TABLE 21.3 : **Results from biofertiliser trials on Rice 1993-94**

(q/ha)

Location	Absolute Control (A)	(A)+R.S. (B)	Recommended fert. dose (C)	(C)+L.S. (D)	(C)+R.S. (E)
Allahabad	36.50	41.80	41.00	-	44.83
Burdwan	40.25	41.00	56.75	57.00	58.75
Nellore	33.39	36.38	61.44	61.64	61.95
Mean	36.71	39.72	53.06	-	55.17

situations viz., microbial population of effective strain, soil pH, crop cultivation environment etc. This necessitate identifying effective strains for specific areas and then promoting the use of biofertilisers. Generalised programme may not create desirable impression and impact on farming community. IFFCO has initiated a special project on promotion of biofertilisers in identified blocks and districts of West Bengal. The coverage of the programme is highlighted below:

District	Block	Crops	Biofert.	Area coverage in acres	No. of culture packets
Malda	Habibpur Manickchuk	Pulses	Rhyzobium	4000	12000
Birbhum	Mayureshwar II	Rice	Bacilum	2000	8000
South 24 Pargana	Joynagar I	Rice	Azolla	100	7000
East Midnapur	Egra I Egra II	Rice	Azotobacter	2000	8000

The programme has been initiated since Kharif 95 for a period of three years. The programme is being implemented in collaboration with Department of Agriculture, Government of West Bengal, Bidhanchandra Krishi Vishwavidyalaya, Kalyani and local NGOs.

Integrated Plant Nutrient Supply System

Biofertilisers are an integral part of the IPNS programme. Use of Rhizobium, Azotobacter, Azosprillum, PSB etc., wherever required have been incorporated in the IFFCO's IPNS programme.

Observation plots have been laid out to recognise the contribution of various nutrient sources in nutrient input-ouput status under specific crop rotations and to workout the nutrient balance sheets of the observation plots and also of the whole village. Cropping pattern have also been changed at few locations to accommodate pulses and introduce Rhizobium inoculation.

Enhancing Biofertiliser Production

Availability of effective strains through an efficient marketing network is quite crucial in biofertiliser promotion. A good number of fertiliser industry members, including IFFCO, have initiated production of biofertilisers by establishing their own biofertiliser production unit. IFFCO has set up a biofertiliser production unit with a capacity to produce 75 MT biofertilisers at Phulpur (Allahabad), U.P. Financial assistance has been obtained from Government of India to establish this plant under National Project on development and use of Biofertilisers. The unit is likely to commence production from the first quarter of 1996.

Summary and Conclusion

Use of biofertilisers have been adequately recognised to supplement the nutrients availability particularly nitrogen and also to enhance the phosphorus availability through solubilisation. Since biofertiliser strains are specific to crops and soils, it is desirable to delineate areas of effective strains and take up extension programmes with intensive area approach. Besides engaging media like farmers meeting, field days, crop literature etc., special efforts through field demonstrations/trials, minikit distribution, IPNS and biofertiliser productions have been undertaken by IFFCO for promotion of biofertilisers. Similar strategy can be adopted by others interested biofertiliser promotion.

PART NINE

INDUSTRIAL AND URBAN WASTE

22

MANAGEMENT OF MUNICIPAL SOLID WASTE

—PAWAN SIKKA

The subject of management of MSW has assumed greater importance today in view of the mushrooming development of urbanisation and industrial growth. The difficulties in finding additional land for dumping of garbage have posed challenging issues and have made people to consider alternative ways for treating the solid waste. The scientists and technologists have now come to appreciate the need for a total approach for waste management in which the urban waste—solid and liquid is classified into different components. For each of the components, suitable technologies could be considered for the treatment and conversion into materials with economic value.

CATEGORIES OF WASTE

The urban garbage can be divided into bio-degradable and non-degradable parts. The bio-degradable part of the garbage can yield organic fertilisers either through anaerobic fermentation in a bio-digester or by aerobic process. The gas produced in the anaerobic process can be used for generation of power or as fuel by households. Bio fertilisers are expected to be sold commercially.

The non-degradable part of the garbage can also be considered in two parts, recyclable materials as distinguished from the non-recyclable ones. We have already noted that the recyclable materials are reprocessed by the industries. The part of the non-degradable garbage which cannot be re-processed can further be subdivided into combustible and non-combustible components. Both of these components can be pelletized under pressure to form combustible and non-combustible types of pellets. The combustible fuel pellets can be used by households or in burners for incineration, which would be necessary for the surgical wastes. The remaining non combustible part of the garbage can be compacted under pressure to form bricks or tiles which can be used for low cost building or for paving the roads.

RECYCLING TECHNOLOGIES

The approach of the new technologies for recycling holds out hope for their application to urban wastes. Various alternative processes for treatment and disposal of urban solid waste may be classified into the following groups:

1. Physical Processes
   (a) Pelletisation
2. Bio-chemical processes
   (a) Aerobic composting
   (b) Anaerobic digestion
   (c) Vermiculture
3. Thermal Processes
   (a) Incineration
   (b) Gasification
4. Recycling
   (a) Plasitc, rubber, glass, iron scraps etc.
5. Miscellaneous
   (a) Dumping of stones, dead animals etc.

A. Sanitary Land-Fill

Though availability of land for disposal of municipal wastes is becoming scarce day by day, yet, since Sanitary landfill is by far the cheapest method of waste management, it has to be a main and major component of such an integrated scheme. Such scientific Sanitary Landfill should have proper lining underneath, intermediate earth covers and probes to arrest methane gas generated in the landfill. Two particular effects normally create nuisance in landfills if not properly controlled properly—first, bad odour and secondly, potentially toxic liquids called leachates that percolate underground to contaminate aquifers, rivers, lakes or oceans. Sanitary landfills are lined underground by dense clays and often heavy plastics. In some, methane vents are set up like tall chimneys before the refuse is dumped around them. A lattice of pipes at the bottom collects the leachates and pumps them to a treatment centre.

B. Incineration

Surgical wastes from hospitals are a major source of health hazard in the city. In the present method of mixed dumping of municipal refuse these hospital wastes pollute the environment and also leaches underground to contaminate the ground water. Moreover, so many rag-pickers who operate on the garbage come in contact of such hazardous wastes and get infected. Hence the only way out to dispose of these surgical wastes is to incinerate them. The incinerator is to be operated on dual fuels—electricity/oil fired and refuse derived fuel (RDF) pellets to be obtained from the RDF plants in the same scheme. Trucks specifically earmarked for the purpose will collect wastes from major hospitals and nursing homes and feed them directly to the incinerator. The cost of incineration can be recovered from the charges levied on the hospitals for services rendered. The hot flue gas can be used in the bio-digesters to accelerate the digestion process. The incineration as such does not generate any marketable product of by-product. This has to be considered as a social obligation and the environmental cost assigned to its usefulness can hardly be overestimated.

C. Refuse Derived Fuel Plant (RDF) Pelletisation

Combustible components of the waste materials are sorted and shredded to denser and consistent form mixed with sawdust and agro/refinery wastes to attain high calorific value. This is pressed to pellets to serve as solid fuel. This pellet can be used, as per requirement, in the incineration described above and the balance quantity can be sold in the local market. This definitely will have a profitable consumer market provided it can be promoted properly. Indigenous technologies are available and this method is a already under trial in Bangalore, Baroda and Bombay.

D. Anaerobic Digestion

In this biological process, the organic bio-degradable portion of the waste, in the absence of oxygen, is converted into methane gas. This also yields manure of high fertiliser quality and in the form of slurry. In down-stream processing this slurry can be used in production of bio-fertilizer or in Spirullina culture. The gas so produced can be used a fuel or in cogeneration of electricity. The main slaughter house in Calcutta generates about 8 MT of highly bio-degradable waste a day, which can be collected separately and fed directly to the Bio-gas plant. The hot fuel gas from the incinerator can effectively be used in accelerating the digestion process in the bio-digesters. Such bio-gas plants are expected to replenish most of their running cost through sale of power.

E. Vermiculture

In this process, earthworms are cultivated and through them the organic portion of the solid waste is converted into very high value manure. This is a new low-cost and environmentally benign technology that requires commercial scale trial. Application of this method in Bangalore and Pune have reportedly been successful.

F. Low-Cost Building/Construction Materials Production Unit

Using the non-combustible and non degradable component of the municipal wastes, a unit is proposed to be set up for production of low-cost building or construction materials such as tiles, bricks, coarse aggregates, pavement blocks etc.

INTEGRATED APPROACH FOR MANAGEMENT OF MSW

It is suggested that—

(i) a coordinated strategy should be chosen to document all the technology—adopted and proven successful at various places in the country and replicate the tried and tested ones under one umbrella;
(ii) municipal corporations should provide free land for the management of MSW to the NGOs and other interested

organisations, preferably in the dumpyard zones for processing the garbage under an integrated system;

(iii) financial institutions, industrial houses, Planning Commission besides the Ministry of Environment and Forests, Ministry of Non-Conventional Energy Sources, Ministry of Rural Development besides the World Bank and UNEP should provide subsidy and financial resources for taking up the MSW project on a national level in the country.

The disposal and utilisation of municipal solid waste is a vital problem and the adoption of an integrated approach for its management would not only provide manure, energy, employment and recycled products but would yield social and economic benefits to the society. It relates to the attainment of a pollution-free atmosphere—the value of which is difficult to estimate. Let us manage the MSW with full vigour and dedication.

PART TEN

DRAUGHT ANIMAL

23

STATUS OF RESEARCH AND DEVELOPMENT ON DRAUGHT ANIMAL POWER IN INDIA

—N.S.L. SRIVASTAVA

Traditionally and even today, notwithstanding the increasing mechanisation of Indian agriculture, working animals, particularly bullocks, continue to be the predominant source of energy for traction and rural transport in India. Year after year, cows replace and add to the required number of working bullocks which has been their primary function. Today, a vast herd of approximately 73 million bullocks and about 8 million male buffaloes provide approximately 32,000 MW of power and help cultivate about 90 million farm holdings which accounts to approximately 63% of the total cultivated area in India, besides being the back-bone of rural transport. At an optimal replacement rate of 10 bullocks per tractor, this herd is equal to about 8 million tractors, thus saving the nation about 26 million tonnes of diesel annually, almost equal to the total production of petroleum in the country, worth Rs. 22,000 crores at the prevailing market price of diesel oil. The market value of draught animals (bullocks, buffaloes, camels, horses, mules, donkeys) in the country is estimated as Rs. 34,500 crores and of animals drawn implements and carts as Rs. 24,000 crores. The replacement value of these draught animals, implements and carts by tractor system will require an investment of approximately Rs. 160,000 crores.

In addition to this the draught animals provide about 100 million tonnes of dry dung estimated approximately Rs. 2,000 crores annually and raw materials for 1.5 million biogas plants in the country saving about 5 million tonnes of fire wood at an estimated cost of Rs. 500 crores annually.

Besides contribution to food and energy, a whole lot of byproducts—hide, skins, bone, hoof, horn, tallow and so many other uses in pharmaceutical industry—are obtained from the fallen and slaughtered animals. There is lack of data on these aspects.

Present Status of Research on DAP

To increase the utilization of animal energy with enhanced man-animal-machine system efficiency, the Indian Council of Agricultural Research is funding number of research projects and schemes. The present status of work done by ICAR centres and other organisation on different aspects of draught animal power are summarized below.

Studies on Draughtability of Animals

The draught power of an animal depends on the species, breed, sex, size, body weight, nutrition and health, environment, its training for work and terrain conditions. As a thumb rule it is taken that an oxen can exert draught force equivalent to 10% of its body weight in sustained working on the whole day basis.

Systematic studies conducted on draughtability of different breeds of bullocks, buffaloes, camels and donkeys in sustained working have shown that:

(a) Malvi (CIAE, Bhopal, 1990-93), Nagouri (KVK, Rewari, 1990-93), Khillari (CAE, Raichur, 1990-93) and Hariana (AAI-Allahabad, 1990-93) breeds of bullocks were able to exert draught, in sustained working (7 to 8 h/d in 2 sessions) equivalent to 12% of their body weight during summer and 14% during winter using local yoke. With improved 3-padded collar harness they were able to exert 14% and 16% load during summer and winter seasons, respectively.

(b) Studies on crossbred humpless bullocks Red Dane and Sahiwal (Rautray and Srivastava, 1982). Jersey and Sindhi and a few other crossbreds have shown that crossbred bullocks are good for draught purposes. Rautray and Srivastava (1982) reported that the crossbred of Red Dane and Sahiwal could work in sustained working for tillage operation during summer up to draught load equivalent to

12.9% of their body weight. Maurya and Devdattam (1982) and Maurya and Guruswamy (1981) also reported similar results on Jersey and Sindhi crossbred bullocks.

(c) He-buffaloes (GBPUAT, Pantnagar, 1990-93) were able to exert draught in sustained working of 6 h in 2 session equivalent to 12% of their body weight using local yoke in both summer and winter seasons.

(d) Camels (CTAE, Udaipur, 1990-93) were able to exert draught load in sustained working of 7 h (in 2 sessions) equivalent to 18% of their body weight. However, following 2 h work in 2 h rest schedule they could work for 6 h at draught load equivalent to 26% of their body weight.

(e) Donkeys (CTAE, Udaipur, 1990-93 and CAE, Raichur, 1990-93) were able to exert draught load in sustained working for 6 hour (in sessions) equivalent to 32% of their body weight. However, they could work up to 36% draught load for 4 h and 40% for 1 to 2 h.

Improved Yokes and Harnesses

The draught force from the animal to the implements is transmitted through yokes and harnesses. How much draught the animal will exert depends on the yoke or the harness used.

Improved wooden yokes and single and double animal harnesses with padded collars have been developed for bullocks and he-buffaloes. Swamy Rao (1964) reported that Allahabad single and double animal collar harnesses gave about 59% higher power output as compared to local yokes and 38% as compared to Nagpuri yoke. In the co-ordinated trials (CIAE, Bhopal, 1993; Singh, 1985, Srivastava, 1994) conducted on the Nagpuri wooden yoke, 7-15% higher power output was obtained while Allahadad type single and double animal harness with 3-padded collar gave 15-25% more power output in comparison to local yoke under sustained working.

Nagpuri yoke and modified Allahabad type 3-padded (CIAE, Bhopal, 1990-93) collar harness were extensively evaluated in farmers fields. The advantage of collar harness in terms of output of work was more pronounced when the animals worked in higher draught range and when draught load was more than 12% of the body weight, i.e., when the animal had to work in the discomfort zone with local yoke.

At GBPUAT, Pantnagar adjustable collar yoke and collar harness for buffaloes have been developed which could be used for different crop row spacings.

Improved Animal-Drawn implements and Carts

A large number of designs of ploughs, harrows, pudders, cultivators, seed-cum-fertilizer drills, weeders, diggers, patela harrow, bund formers, multi-purpose tool carriers, etc., were developed at CIAE, ICAR Co-ordinated Research Scheme on Farm Implements and Machinery and Agricultural Universities and are commercially available and used by the farmers. These implements give greater coverage as compared to traditional implements.

A large number of animal-drawn processing equipments like chaff cutter, cane crusher and oil-extracting machines are also commercially available. Animal-drawn 'Rahat' and duplex pumps for water lifting are also used. The CIAE, Bhopal (Shanmughaum, 1992) and CAPART have brought out a number of monographs and bulletins giving details of different designs of animal-drawn equipment.

Lot of work has been done on the design of the bullock-carts. Improvements have been made mainly to replace wooden axle by iron axle providing ball or taper roller bearing instead of bushes, replacing wooden wheels by pneumatic wheels or providing iron or rubber rims on wooden wheels, providing punctureless tyres, FRP body and better designing of chassis on scientific basis for higher pay loads and human comforts. These improved carts can carry loads from 1.5 to 3T as compared to 0.8 to 1 T load carried by local bullockcarts.

CIAE, Bhopal (Deshpande and Ojha, 1984) and IIM, Bangalore (Ramaswamy and Narsimhan, 1984) and many other organisations have brought out publications giving specifications and capacities of various designs of animal carts.

Matching of Implements to Draught Animals

At CIAE, Bhopal (1990-93), use of draught animals with matching implements for tillage, sowing and weeding gave greater area coverage, higher power output, lower specific energy consumed for a particular operation and lower cost of operation in comparison to existing sizes of implements. The increase in output obtained ranged from 30 to 70%.

Annual Utilization of Draught Animals

With the modernisation of agriculture and availability of electricity and metalled road system in the villages, the use of draught animals for water lifting, cane crushing and rural transport has gone down. The present day annual utilization in different regions vary from 300 h to 1500 h whereas under ideal situations, the animals sould be used for about 2400 h annually. The annual utilization of DAP can be increased by increasing use of DAP in rotary mode of operation for agro-processing work and operating new machines for which they are not being used presently.

Use of Draught Animals in Rotary Mode

During idle period, the draught animals can be used in rotary mode for various agro-processing operations like chaff cutting, flour grinding, soy flaking, grain cleaning, threshing, maize shelling and decorticating. The output of such machines are under:

Flour grinder	8-25 kg/h
Soy flaking machine	23-25 kg/h
Grain cleaner	120-450 kg/h
Paddy thresher	238-850 kg/h
Maize sheller	205 kg/h
Groundnut decorticator	100-150 kg/h
Chaff cutter	380-430 kg green fodder/h
	100-150 kg dry fodder/h
Paddy cleaner	710-1,100 kg/h
Duplex pump	1,800 lt/h at 1m head

Such agro-processing plants would be quite suitable in rural areas.

In certain parts of Bihar, Orissa and Karnataka, cows having low milk yield are also used for draught purposes.

Use of Pack Animals for Draught Purposes

As per 1977 Census, the population of pack animals (donkeys, mules, horses, camels and yaks) in India was approximately 3.2 million. These pack animals are mainly used for carrying loads on the back. A study conducted at Udaipur (CTAE, Udaipur, 1990-93) showed that camels could carry a packload of 50 kg continuously for 5 h, 100 kg for 4 h, 150 kg for 3 to 5 h and 200 kg for 3 h on sandy track while they can carry about 1,600 kg load in a cart on sandy track and 2,000 kg on tar macadam and kuchcha roads.

Similarly donkeys (CTAE, Udaipur 1990; CAE Raichur 1990-93) could carry a pack load of 20-30 kg for 2-3 h, 50 kg for 2 h and 70 kg for 1 h while in a donkey cart they can pull load even up to 800-1,000 kg on tar road.

Quantifying Fatigue Symptoms and Scheduling Work-Rest Cycle

Upadhyay (1987) prepared a fatigue score card which is useful and is currently being used for studies on bullocks and buffaloes. Fatigue score cards have also been developed for camels and donkeys at CTAE, Udaipur.

Based on the studies conducted at Udaipur, Rewari, Pantnagar, Allahabad and Raichur, the following work-rest schedule was found better from the work output point of view:

Bullocks

(i) 2 h work + 1 h rest + 2 h work + 1 h rest + 2 h work
(ii) 3 h work + 1 h rest + 3 h work
(iii) 4 h work + 2 h rest + 3 h work

Buffaloes

(i) 2 h work in the morning + 7 h rest + 4 h work in the evening
(ii) 2 h work + 1 h rest + 2 h work + 1 h rest + 2 h work + 1 h rest + 2 h work in continuation.

Camels

2 h work + 1 h rest + 2 h work + 1 h rest + 2 h work + 1 h rest + 2 h work

Donkeys

1 h work + 1 h rest + 1 h work + 1 h rest + 1 h work

Instrumentation

Loading Cars

Several types of loading cars have been developed in India for draughtability studies. They are (i) sledge type, (ii) winch type, (iii) suspended weight type, and (iv) hydraulic type.

The CIAE animal loading car (Srivastava and Pandey, 1992) of 300 and 500 kgf draft capacities is best from the point of ruggedness, versatility and reliability. Both these loading cars are similar in design and are of hydraulic type. A hydraulic pump is driven off the ground wheels. Braking effect on the wheels is created by restricting the flow of the hydraulic oil, thus causing load on the animals. Maintenance of load settings is not affected by variation in speed in the range of 1.8-5 kmph.

Dynamometers for Measuring Pull

Spring, hydraulic and load cell type dynamometers are commercially available for measuring pull. Load cells are more precise and accurate and are preferred over the other two types of dynamometers. Suitable fixtures have been developed for proper mounting of these dynamometers on implements.

Measuring of Angle of Pull

For measuring angle of pull an Abney level should be used. This level is kept on the beam or rope and indicate the angle of pull when the spirit level provided in the instruments is set to a horizontal position.

System for Measurement Under Field Conditions

To develop an animal performance logger (O'Neill, *et al.*, 1987), instruments have been developed and evaluated under a collaborative programme between AFRC Institute of Engineering Research, Silsoe (UK), and CIAE, Bhopal (India), for measuring physiological responses of draught animals while working in the field Simultaneous measurements of heart rate, respiration rate, body temperature, stepping rate, pull, angle of pull and speed of working are done using suitable sensors. These sensors are fitted on the animals and the implements being used. The leads of all these sensors are connected to a junction box fitted on the yoke. From this a common lead goes to a backpack battery-operated signal conditioning unit and to a logger. A battery-operated microcomputer is used to analyse the data.

The field data were analysed within 1-2 h of test run and print outs of statistically analysed data were obtained. The unit is to be commercialised.

REFERENCES

AAI, Allahabad. 1990-93. Annual Report Allahabad Centre of AICRP on Utilization of Animal Energy, AAI, Allahabad.

CAE, Raichur. 1990-93. Annual Report Raichur Centre of AICRP on Utilization of Animal Energy, CAE, Raichur.

CIAE, Bhopal. 1990-93. Annual Report Bhopal Centre of AICRP on Utilization of Animal Energy, CIAE, Bhopal.

CIAE, Bhopal. 1993. Annual Report and Proceedings of Fourth Annual Workshop 1993. AICRP on Utilization of Animal Energy, CIAE, Bhopal.

CATE, Udaipur. 1990-93. Annual Report, Udaipur Centre of AICRP on Utilization of Animal Energy, CTAE, Udaipur.

Deshpande, S.D. and Ojha, T.P. 1984. Traditional and Improved Bullock Carts in India. Technical Monograph No. 44, CIAE, Bhopal.

GBPUAT, Pantnagar. 1990-93. Annual Report Pantnagar Centre of AICRP on Utilization of Animal Energy, GBPUAT, Pantnagar.

KVK, Rewari. 1990-93. Annual Report Rewari Centre of AICRP on Utilization of Animal Energy, KVK, Rewari.

Maurya, N.L. and Devdattam, D.S.K. 1982. Work Performance of crossbred bullocks. *Indian Journal of Dairy Science* 35(2).

Maurya, N.L. and Guruswamy, T. 1981. Draught characteristics of crossbred and local bullocks. A case study. *Curr. Res.*10(4) : 4-6.

O. Neill, D.H., Howel, P.J.L., Paice, M.E.R. and Kemp. D.C. 1987. An instrumentation system to measure the performance of draught animals at work. In Utilization and Economics of DAP. (Eds) Srivastava, N.S.L. and Ojha, T.P., CIAE, Bhopal.

Ramaswamy, N.S. and Narsimhan, C.L. 1984. Indian Animal-drawn Vehicle. IIM, Bangalore.

Rautaray, S.K. and Srivastava, N.S.L. 1982. Field performance of crossbred bullocks in black soils during tillage operations in summer. Proceedings of the National Seminar on Draught Animal Power System in India. IIM, Bangalore, July 16-17, pp. 677-99.

Shanmughum, C.R. 1982. Farm Machinery and Energy Research in India. Technical Monograph No. 32. CIAE, Bhopal.

Singh, V.V. 1985. Development and evaluation of single and double animal harnesses. Annual Report of Utilization and Economics of Animal Energy Project, CIAE, Bhopal.

Srivastava, N.S.L. and Ojha, T.P. 1987. Utilization and Economics of Draught Animal Power. Technical Bulletin No. 51. CIAE, Bhopal.

Srivastava, N.S.L. 1989. Research on draught animal power in India. In : Draught Animals in Rural Development, Proceedings of the International Research Symposium, Cipanes, Indonesia, (Ed.) Hoffmann, D., Nari, J. and Petheram, R.J. pp. 53-60.

Srivastava, N.S.L. and Pandey, M.M. 1992. Design and development of CIAE animal loading car for draughtability studies. *Indian J. Agric. Engg.* 2(2): 79-84.

Srivastava, N.S.L. 1993. Status of Research on Engineering Aspects of Draught Animal Power. *J. of Rural Energy,* Vol. 2 (2-4) : 15-32.

Srivastava, N.S.L. and Devnani, R.S. 1994. Background information for Regional Conference on Efficient Utilisation of Animal Energy in Rural Development. Held at CIAE, Bhopal on October 21-22.

Swamy Rao, A.A. 1964. Report on Bullock Harness Research Project. Allahabad Agricultural Institute, Allahabad.

Upadhyay, R.C. 1987. Factors limiting work capacity and fatigue assessment in draught animals. In : *Utilization and Economics of Draught Animal.* (Eds) Srivastava N.S.L. and Ojha, T.P. CIAE, Bhopal.

PART ELEVEN

IMPROVED COOKING DEVICES

24

THERMAL PERFORMANCE OF A SINGLE POT CHIMNEYLESS MUD COOKSTOVE

—S.K. SHARMA, R.K. WANCHOO AND SUNEETA CHOPRA

Energy availability problem is becoming more and more acute with every passing day. If urgent measures are not taken it will cause a major setback to the economic and technological growth of the society. Primary needs of the mankind is food and fuel is essential to prepare food. Ever since the time primitive man discovered fire, he has been cooking food in the simple traditional cookstoves (chulha), using the most dominant fuel "wood". This scenario has not changed, for the major part of the world, especially in the rural areas of the developing countries, where about 80 percent of the population still uses traditional fuels like wood, agricultural residues, burnt in traditional cookstoves, for meeting its cooking, heating and other household needs.

Method of Construction

Clay was mixed with sand (sieved through a 4 mm screen). To this mixture, wheat husk and cowdung were added. The ratio of clay : sand : wheat husk : cowdung was 40 : 10 : 1 : 4. The ingredients were mixed thoroughly and kneaded after adding water, as per requirement. The mixture was allowed to ferment for 3-4 days. Additional amount of water, if needed was added to ensure that the consistency of the mixture was such that it did not slump. The above said mixture was filled in a hollow tin mould. After the mud had set, the tin was removed. The outline of the fire box, pot hole and ash outlet was made and the extra clay was carved off. The inner dia of the port hole was 18 cm. and outer dia was 26 cm. The size of the fuel gate was 8 cm. × 7 cm. and the size of the ash outlet was 7 cm. × 7 cm. The height of the chulha was 33.5 cm. A cast iron grate was provided for proper combustion. The detailed dimensions of the cookstoves are shown in Fig 24.1.

Experimental Procedure

The complete thermal analysis of a single pot, Naranjan-I cookstove was undertaken at four different burning rates varying from 0.5 kg/h to 1.25 kg/h with successive increments of 0.25 kg/h each. The testing was done for a period of two hours using standard water boiling test[1,2,3]. Standard calibrated thermocouples were used for measuring the temperature of the water, pots, stove walls, flue gas and flame. The composition of the various components of the flue gases such as carbon monoxide, carbon dioxide and oxygen were measured during the testing period using an on line gas analyser (Rekin Keiki).

After the expiry of the burning period i.e., when the flame has died down, the pot was replaced with fresh pre-weighed pot containing water. The temperature of the water, pots, stove walls and flue gas were recorded, till the water in the pot showed a fall in temperature. Charcoal and ash were weighed on an analytical balance.

Results and Discussion

The efficiency of Naranjan-I cookstove at different wood burning rates has been presented in Table 24.1. The results indicates that an efficiency of 22.32% was attained at a wood burning rate of 1.00 kg/h, which was the main design criteria. A fall in efficiency at a wood burning rate of 1.25 kg/h was noted. This may be due to the decrease in heating rate with the increase in burning rate beyond 1 kg/hr as shown in Fig 24.3. This could be due to back draft, as a result of the inability of the flue gas outlets to handle the flue gases generated during the combustion process at higher burning rate. This has been purposely done so as to give indication to the user that the fuel feeding need to be reduced.

The time taken to boil 5.00 litres of water contents at different wood burning rates is given in Fig 24.2. The results indicate that the

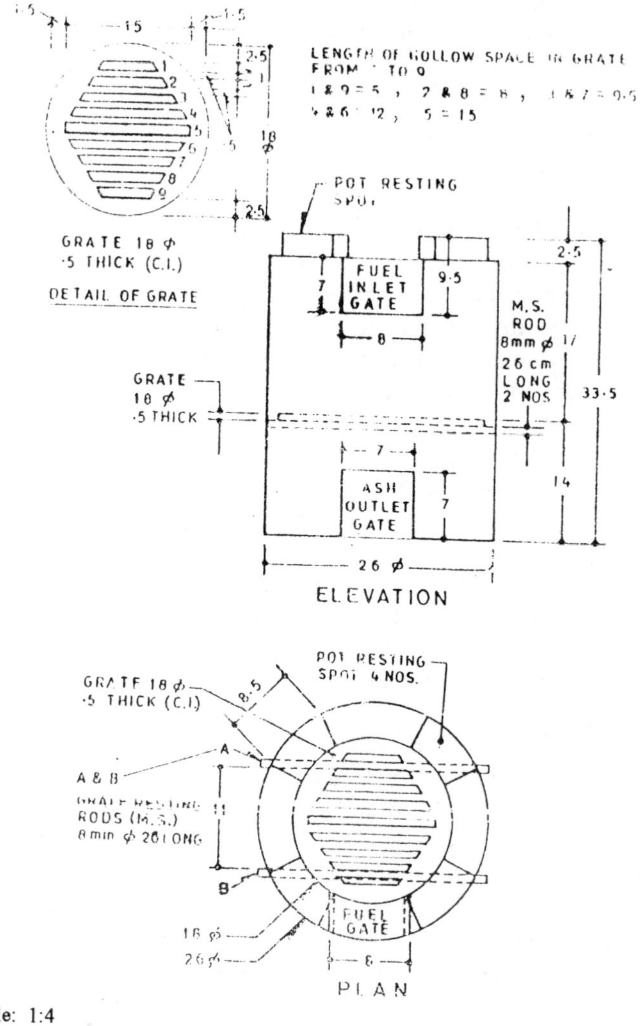

Scale: 1:4
Dimensions are in cm

Developed by
TBSU: At Energy Research
Centre P.U. Chandigarh

Fig. 24.1: Naranjan-I Mud (Chimneyless) Chullah

time taken to boil the water contents in the pot decreases with increase in wood burning rates. This is due to the decrease in the heating rate at higher burning rates as explained earlier. The data is well represented by the following equation:

$$Tb = 82.75 - 79Rb + 28Rb^2$$

Tb = boiling time in (min)

Rb = burning rate in (kg/hr)

From the above equation, boiling time corresponding to any wood burning rate can be evaluated. The variation in pot heating rate with wood burning rate is presented in Fig 24.3. It can be inferred from this figure that the pot heating rate increases as a result of increase in heat flux with the increase in wood burning rate. The data can be represented by the equation:

$$Hr = 0.1374 + .3184\,Rb$$

Hr = rate of heating in °C/(min)(lit), Rb = burning rate in (kg/hr)

From this equation, pot heating rate corresponding to any wood burning rate can be calculated. Although, the rate of heating increases and time taken to boil five litres of water gets reduced at higher burning rate, however, the efficiency of the cookstove decreases with the increase in burning rate as shown in Table 24.1. The main reason for the decrease in the thermal efficiency with increase in burning rate is the increase in various losses as shown in Fig 24.4. It can be inferred from Fig 24.4 that the radiation losses increase with increase in wood burning rates. This may be due to the higher temperature attained as a result of higher energy input as well as increase in the flame temperature with increase in wood burning rate. The convection losses also increase with increase in wood burning rates which may be due to the higher temperatures attained as a result of higher gross energy input. In terms of percentage losses, the radiation losses vary between 10-11 percent and convection losses vary between 2.0-2.22 percent.

In can further be inferred from Fig 24.4 that flue gas losses also increase with increase in wood burning rates as the volume of flue gas increases with increase in wood burning rates. This could be due to the increase in flue gas temperature as shown in Table 24.2.

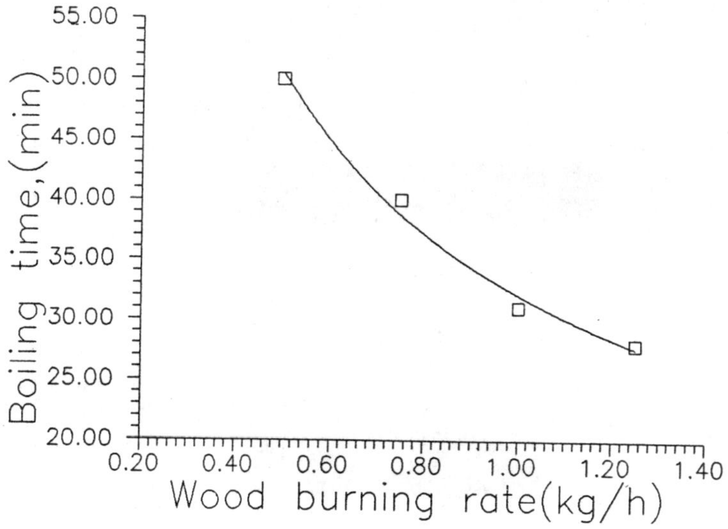

Fig. 24.2: Wood Burning Rate Vs Boiling Time

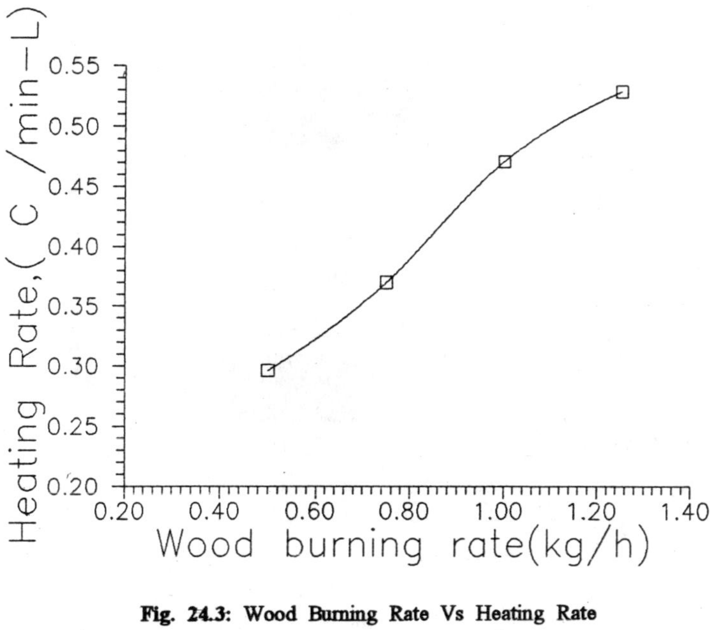

Fig. 24.3: Wood Burning Rate Vs Heating Rate

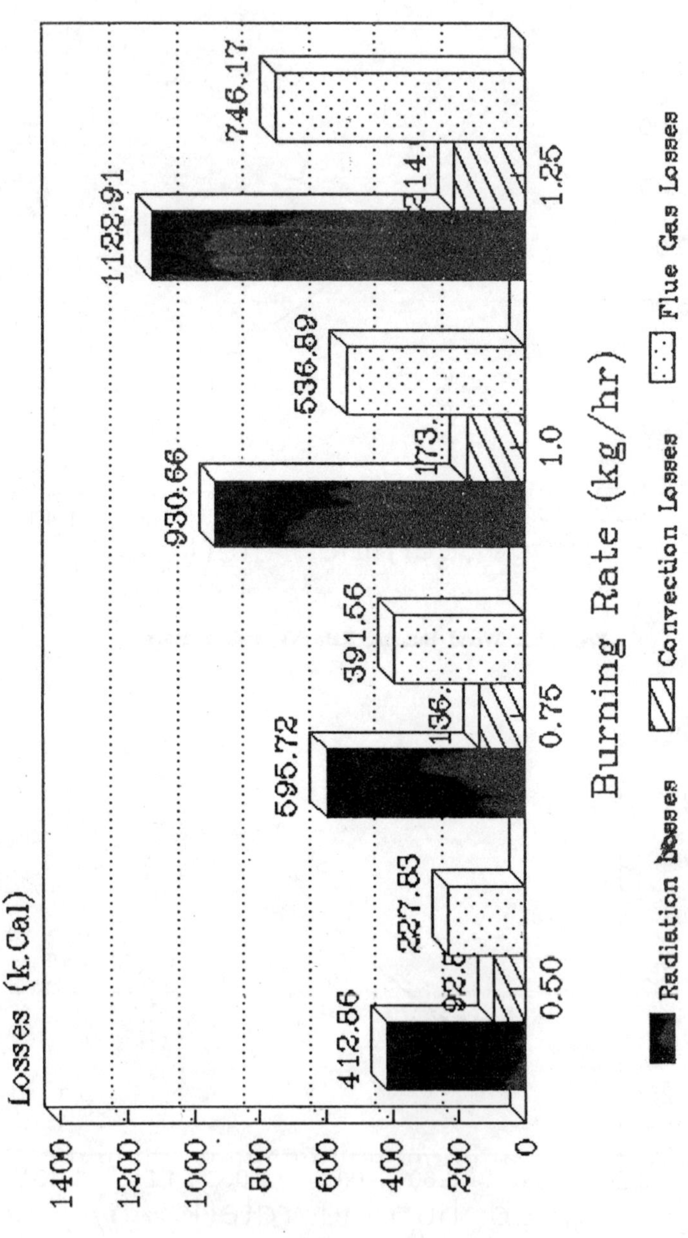

Fig. 24.4: Burning Rate Vs Radiation, Convection & Flue Gas Losses

TABLE 24.1: **Efficiency vs Burning Rates**

Wood Burning Rate kg/h.	Efficiency %	
	Burning Period	Recovery Period
0.50	17.55	20.22
0.75	18.64	20.29
1.00	20.67	22.32
1.25	18.13	19.86

TABLE 24.2: **Flue Gas Analysis at Different Wood Burning Rate**

Burning Rate	0.5 Kg/hr	0.75 Kg/hr.	1.0 Kg/hr.	1.25 Kg/hr
CO_2, Vol. %	3.7	4.2	5.0	4.90
CO, Vol. %	0.23	0.16	0.11	0.26
O_2, Vol. %	18.0	17.2	16.60	17.20
N_2, Vol. %	78.07	82.18	78.29	77.64
CO/CO_2	0.062	0.038	0.022	0.053
Flue gas temperature°C	91	98	108	119

Flue gas analysis at different wood burning rates is presented in Table 24.2. CO/CO_2 ratio is minimum at a wood burning rate of 1.00 kg/h. This observation further supports the higher efficiency attained at this burning rate due to better combustion.

Thus, it can be concluded from the above discussion that the Naranjan-I, due to its low cost, will meet the requirements of weaker section of the society in rural areas on economic considerations.

NOTES

1. Sharma, S.K., Sethi, B.P.S., Chopra, Suneeta, Thermal Performance of Rohini-an Improved Wood Cookstove, Energy Conversion & Management 30(4), 409 (1990).
2. Sharma, S.K., Sethi, B.P.S., Chopra, Suneeta Studies on the Performance of the Improved Cookstove—Abhinav, *SESI Journal* 3, 35 (1989).

3. Sharma, S.K., Sethi, B.P.S., Chopra, Suneeta, Thermal Studies of an Improved Wood Cook Stove-Surbhi, in 'Renewable Energy for Rural Development', V.V.N. Kishore, Editor, Proceedings of the National Solar Energy Convention 1988, Dec. 1-3, Hydrabad Tata-McGraw Hill, pp 324-328.
4. Sharma, S.K., Satish Kumar and Chopra, Suneeta, Thermal Performance of Metal Cladded Ceramic Sona Chulha, Paper Presented at National Solar Energy Convention 1992, Dec. 28-30, New Delhi.

25

COOKSTOVE
A Systematic Design

—P.S. RAJPAL , K.S. SHISHODIA AND G.S. SEKHON

In rural areas in India many households still use traditional cookstoves for cooking (Fig. 25.1) These are made from locally available clay mixed with straw, bricks and/or stones. Agricultural crop residue, dung cake, fire wod etc. are generally used fuels. The availability of fire wood has reduced considerably due to depletion of forest cover. In some regions there is acute scarcity of firewood. So there is a need to make the cookstove more energy efficients, so that these consume comparatively less firewood in cooking (Rajpal, 1991).

NEED ANALYSIS

In rural areas of India, traditional cookstoves are mostly used. In the states of U.P., Haryana, Bihar, Punjab, Madhya Pradesh and Rajasthan, where about 44% of total population of India lives, U-type of cookstoves are commonly found. In Bihar two pot hole stoves are also used. Fuel used in these stoves is generally of agricultural crop residue, animal dung cake and fire wood collected from nearby waste land or forest. The efficiency of these stoves is in the range of 10% to 18%. Sometimes coal is also used.

Problem analysis

The inputs, outputs and evaluation criteria relevant to the present design are given below.

(a) Input Variables

1. Raw food : (includes wheat flour, rice, pulses, potatoes,

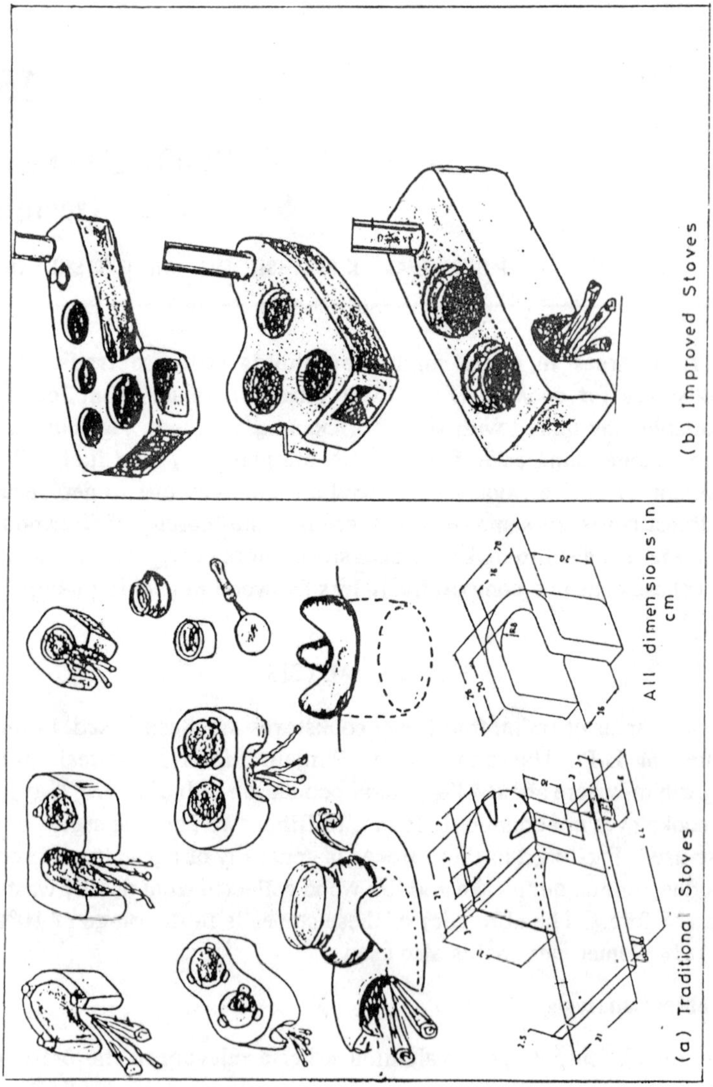

Fig. 25.1: Some Traditional and Improved Cook Stoves

peanuts, onions, cooking oil, water etc.) total about 2.5 kg in about 30 minutes.

2. Fuel : Dung cake of 17 cm to 22 cm dia and 1.5 cm to 2.5cm thickness.
   Fire wood : 1 cm to 10 cm in dia and 10 cm to 80 cm in length.
   Agricultural crop residue : Arhar stalk, corn cob and mustard stalk.
   Ambient air : Composition 21% O_2 by volume Temp. 5°C to 44°C, velocity upto 18 km/hr, relative humidity 19 to 90%.

(b) Output variables

Cooked food, bread, Hot water, smoke, steam escaping into atmosphere during cooking, auxiliary heat from stove body and from fuel inlet, light from combustion chamber, ash.

(c) Evaluation Criteria

Performance	-	Thermal efficiency > 25%
Operation, repair and maintenance	-	Should be better than the traditional stove
Safety	-	No fire hazard
Reliability	-	Better than that of traditional stove
Removal of smoke	-	Should preferably remove smoke from the cooking site
Multifuel flexibility	-	Must perform reasonably well on locally available fuel (S)
Fuel preparation requirement	-	Should not require elaborate fuel repreparation.
Cost to consumer	-	About Rs. 200 with equal amount of Govt. subsidy
Manufacturing effort	-	Simple manufacturing by local artesans/ workshops
Life	-	About three years (or 3300 cycles of operations).

Sub-Problems

The problem of designing a cookstove can be divided into following sub-problems :

(a) Design of combustion chamber, (b) design of grate, (c) design of chapati baking chamber,(d) provisions to extract heat from the exhaust, (e) mechanism to discharge smoke, (f) preparation and feeding of fuel, (g) feeding of air for initiation and sustaining combustion, (h) feeding and placing of food and utensils and (i) material for construction.

Solution Synthesis

Twenty-one alternate designs of the cookstoves were synthesized. A few of these designs are shown in Fig. 25.2 and 25.3. Each design was evaluated from physical realizability, economic viability, financial feasibility and social acceptability considerations. All the solutions were found to be physically realizable and financially feasible. Economic viability transformed in expected cost and a few other features of some of the cookstoves are given in Table 25.1.

Preliminary Design

The different design concepts were ranked on the basis of evaluation criteria mentioned in problem analysis step. The decision process involved the assignation of relative weightages to each criterion and probability ratings of each concept, vis-a-vis each different criterion. Solution No.8, secured the highest ranking. A schematic view of the chosen solution is given in Fig.25.4. It incorporates almost all features required by the average consumers. In additions it has a chamber to keep the cooked food warm. The hot chimney does not create any fire hazard to the thatched roofing as it has been taken out from the wall.

The proposed solution was found satisfactory from compatibility and stability points of view.

Detailed Design

(a) *Stove Design*: Detailed designed of stove (Fig. 25.5) was done as per procedure outlined by Rajpal (1991,1994). The combustion chamber height/diameter ratio was taken with

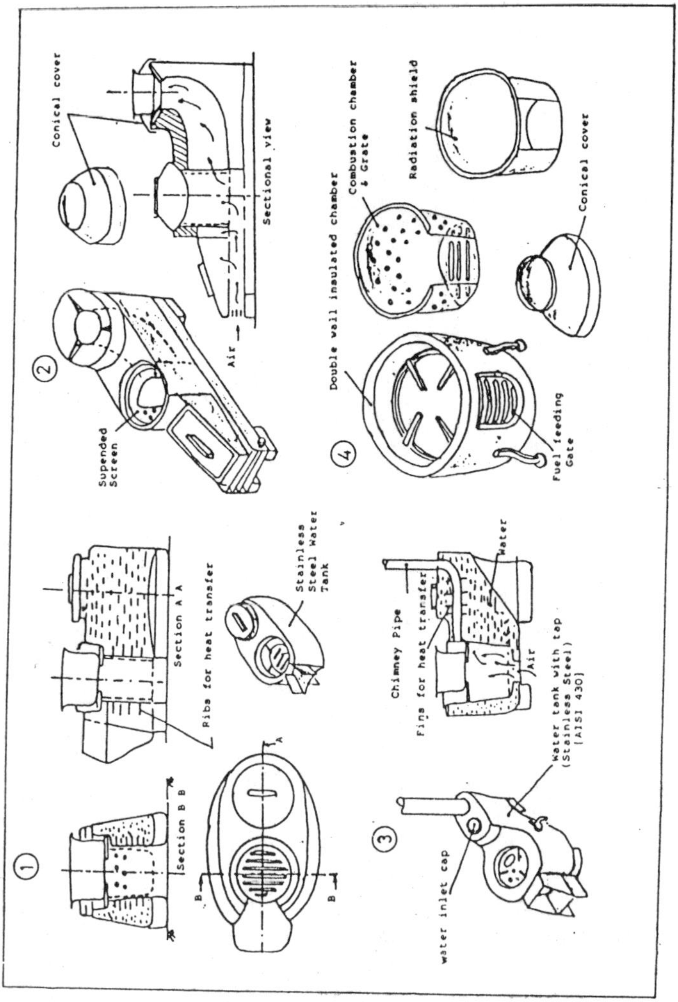

Fig. 25.2: Alternate Designs of Cookstove

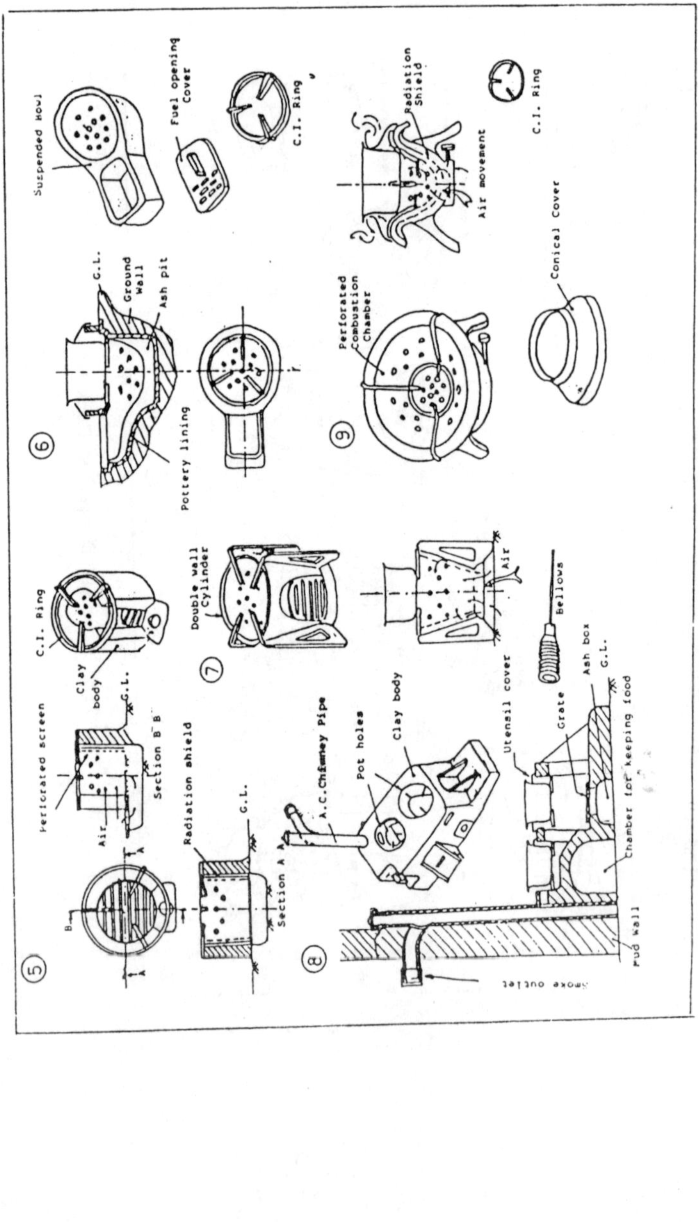

Fig. 25.3: Some More Designs of Cookstove

TABLE 25.1

Solution No.	Anticipated production cost in Rs.	Chapati Baking system	Special features
1.	740	Slightly different baking system	Portable, two pot chimneyless stove stainless steel body
2.	320	-do-	-do-
3.	800	Traditional baking system	Fixed single pot stove with chimney stainless steel body
4.	200	Different baking system	Portable, single pot, chimneyless stove.
5.	165	Traditional baking system	-do-
6.	150	Different baking system	Fixed single pot underground, chimney-less stove
7.	130	Slightly different baking system	Portable, single pot, chimneyless stove
8.	175	-do-	Fixed, two pot, mud stove with chimney and provision for storage of food.
9.	210	Different baking system	Portable, single pot, chimneyless stove.

the help of the empirical formula due to Micuta (1985). Grate area of 15 w/cm^2 is in accordance with the recommendations of De Lepeliere *et al.* (1981). Higher open grate area for burning dung cake has been provided (Maheshwari *et al.*, 1990). Suggestions of Bhatt (1990a, 1990b), Gusain (1990) have been taken into consideration.

(b) *Design of Bellows*: The traditional blow pipe is operated by mouth which is quite an inconvenient method to blow air for initiating and sustaining combustion in the cookstove (Fig. 25.7a). The bellows (Fig. 25.6) was considered a better device for this purpose. It is operated by hand (Fig. 25.7b) and supplies 20 to 30 litres of air per minute when operated

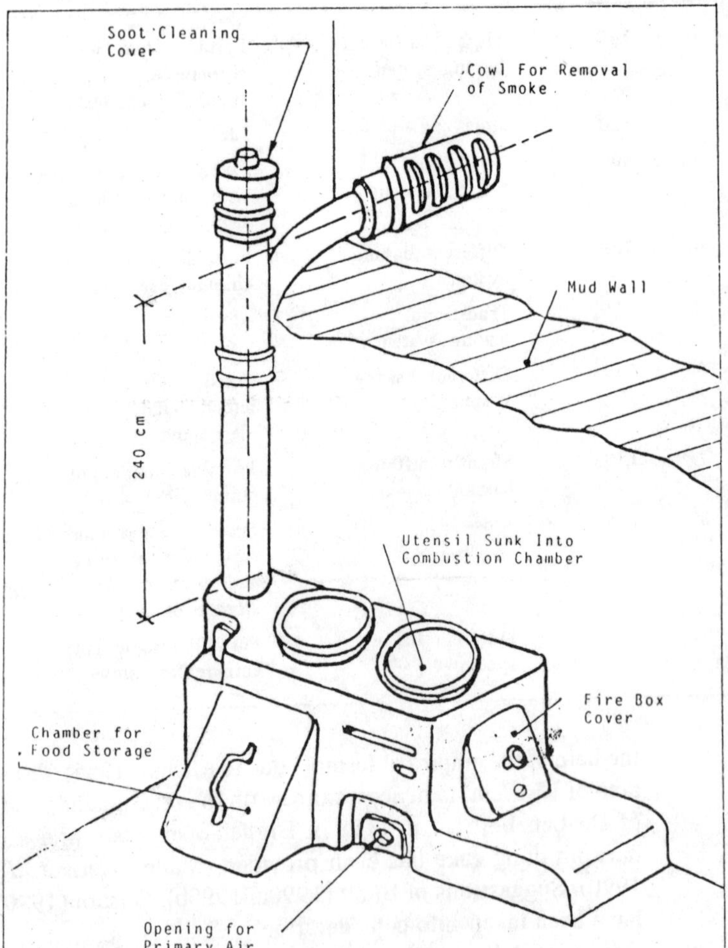

Fig.4 Pictorial View of Cookstove

Fig. 25.4: Pictorial View of Cookstove

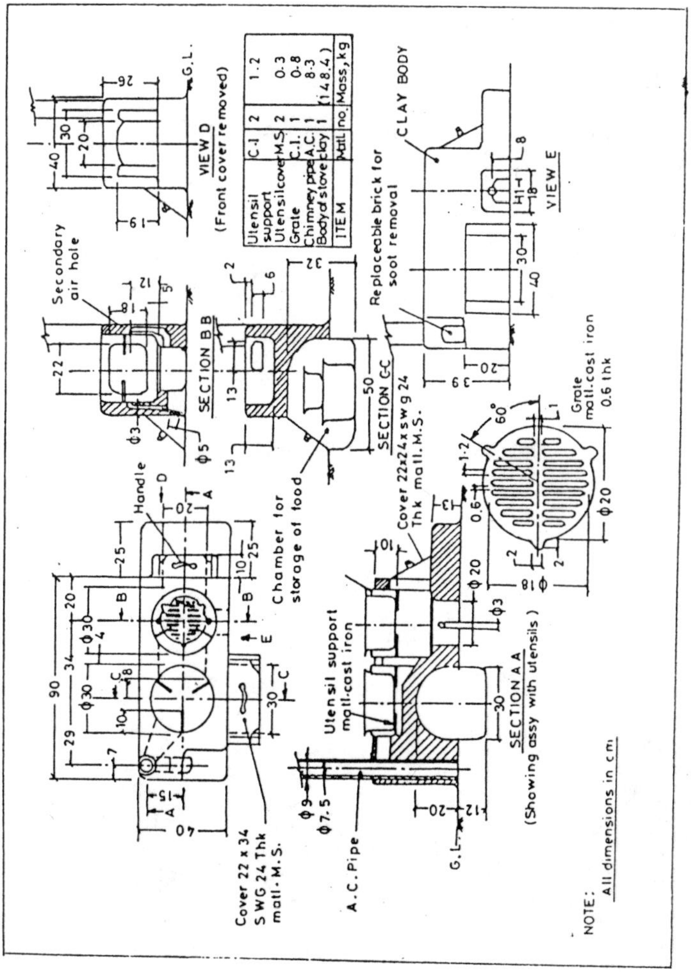

Fig. 25.5: Detailed Design of the Cookstove

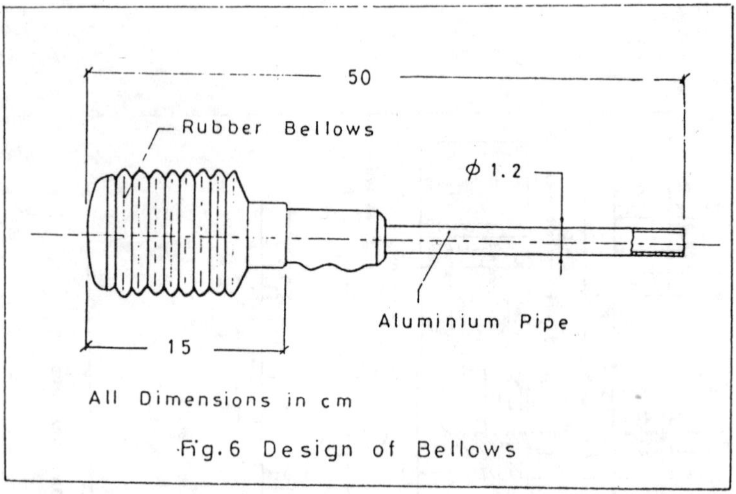

Fig 25.6: Design of Bellows

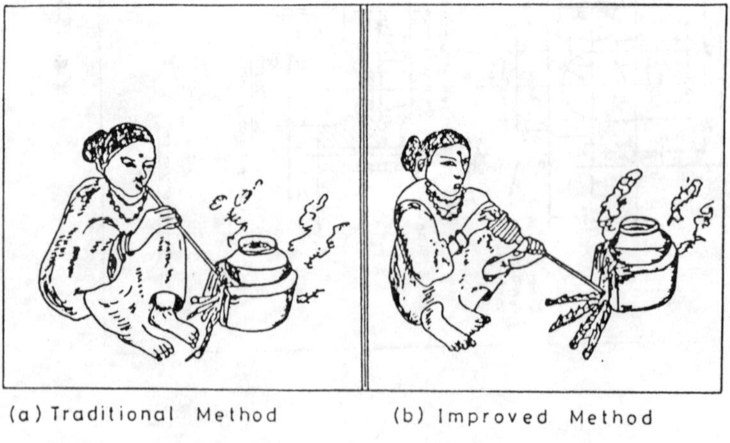

Fig. 25.7: Traditional and Improved Method of Blowing Air

at the rate of 70 to 100 strokes per minute. The estimated cost of bellows is Rs. 35 and it weighs 0.12 Kg.

The stove was compared for its performance with (i) traditional U-type of stove and (ii) Sugam-II stove (Maheswari and Rajpal, 1990).

TABLE 25.2: **Camparison**

Characteristic	Present Design	Sugam II	Traditional U-Type
Thermal efficiency (using wood at burning rate of 1.25 Kg/h) in %	38.1	22.3	14.4
Power output, KW	2.11	1.29	0.83
CO/CO_2 ratio	0.06	0.04	0.15

Cost of Production

The stove requires for fabrication M.S. sheet, Cast iron rings, metal handles, A.C. Chimney pipe, T bend, pottery cap, M.S. fabricated chimney cap. The cost of all these items added to factory, general and sales expenses, plus 10% profit amounts to a total sum of Rs. 175.

Conclusion

(i) The thermal efficiency of the present design is 38.1% which is significantly higher than that of Sugam-II which is 22.3%. The cost of the stove is quite low i.e. Rs. 175 only.

(ii) Smoke is discharged by passing the chimney pipe through the wall using a 'T' joint. This provision allows for easy cleaning of the chimney from inside the kitchen and also avoids contact of hot chimney pipe with thatched roofing.

(iii) The stove has provision for keeping food warm in an enclosed chamber.

(iv) A pair of bellows has been designed as an accessory of the stove.

REFERENCES

Bhatt, M.S. (1990a), "Thermal rating and design of wood stoves: A new approach" *International Journal of Energy Research*, Feb/March, Vol. 14, No.2, pp.169-180.

Bhatt, M. S. (1990b), "Towards the design of high efficiency woodstoves" *International Journal of Energy Research*, August, Vol. 14, No.6, pp.630-649.

De Lepeleire, G., Prasad, K.K., Verhaat, P., and Visser, P.,(1981), A Woodstove Compendium, Eindhoven University of Technology, The Netherlands, pp.307-308.

Gusain, P.P.S. (1990) *Cooking Energy in India*, Har-Anand Publication, New Delhi, pp.15-16, 53-55.

Maheswari, R.C. and Rajpal, P.S. (1990), Annual Progress Report on Cookstoves, Technical backup unit N.P.I.C., Centre RDAT, IIT, New Delhi.

Micuta, W. (1985), Modern stoves for all, Intermediate Technology Publication, London, pp.16-17.

Prasad, K.K., (1983), Wood Heat for cooking, Indian Academy of Sciences, Bangalore, pp.101-103.

Rajpal, P.S., (1991), Design, Fabrication, and Testing of a Cooking Facility for Rural Area, M.Tech. Thesis, Design Engineering, IIT, New Delhi.

Rajpal, P.S., Shishodia, K.S. & Sekhon, G.S., (1994), "Energy Efficient Design of a Cookstove" *International Journal of Energy Sources*, U.K., Vol.16, No.3, July-September, pp. 387-399.

26

EFFICIENT UTILIZATION OF BIOMASS FUELS FOR REDUCING ITS ADVERSE IMPACT ON THE ENVIRONMENT

—R.C. MAHESHWARI AND J.B. KANDPAL

Biomass fuels contribute major fraction of domestic energy requirements in India[1]. While fuelwood and agri-residue fulfils major fraction, animal dung also supply substantial part of above energy requirements[2,3]. Direct combustion of different biofuels releases a number of pollutants which pollutes the indoor environment[5-7]. Among the gaseous pollutants, carbon monoxide, carbon dioxide, nitrogen dioxide and formaldehyde are main[8]. The suspended particulate matter (SPM) and polycyclic aromatic hydrocarbons also causes various health problems. To cope with this problem these fuels may be processed for reducing the high indoor pollutants concentration during various end use applications.

At present also in many parts of India, particularly in small towns and semi-urban areas use of charcoal is preferred as compared to the fuelwood[9,10]. Similarly use of biogas is now promoted instead of direct combustion of dung-cakes.

In the present work an experimental investigation was undertaken to study the emission of various air pollutants from the combustion of fuelwood, dung-cakes, charcoal and biogas in their respective cooking appliances. The air pollutants studied are carbon monoxide, suspended particulate matter and benzo (a) pyrene.

Materials and Methods

The specification of improved mud stove (Sugam-II)[11] used for the combustion of fuelwood and dung-cakes, charcoal stove and biogas stove is given in Table 26.1. The characteristics of four biofuels is given in Table 26.2. The experiments were carried out in simulated conditions in a room of size 2.6m × 2.8m × 2.3m (Room A). This room opens into an another room (Room B) of size 6.6m × 2.8m × 2.3m. The room A has no other arrangement for ventilation. The Room B however, has a window (size 1m × 1m) which was kept open during the experimentation. The investigations were carried out separately for each stove. The stove was placed on the kitchen floor during the experimentation.

TABLE 26.1: **Specifications of the stoves**

Cooking stove		
Sugam-II		
Developed by	:	Centre for Rural Development and Technology, IITD
Size	:	80cm × 36cm × 24cm
Type	:	Fixed Mud
Thermal Efficiency	:	26.5% at 1 kg/hour fuelwood burning rate
Thermal Efficiency	:	25.5% at 1 kg/hour dung-cake burning rate
Charcoal Stove		
Developed by	:	Centre for Rural Development and Technology, Delhi
Fuel used	:	0.56 kg/hour
Thermal Efficiency	:	40.5%
Biogas stove		
KVIC model.		
Thermal Efficiency	:	55%

The charcoal stove was ignited outside in the open area while the other two stoves were ignited in the kitchen itself. The charcoal stove was taken inside once the visible smoke disappeared. Same set of equipment were used for the monitoring of air pollutants for all investigations presented in the work. The air pollutants were monitored at two different heights in the room A (kitchen); (i) at a distance and height of 50 cm from the stove and (ii) at a distance of 50 cm and height 150 cm from the stove. While the height of 50 cm represents the breathing height in squatting posture, the height of 150 cm corresponds to the breathing level of a person standing in the kitchen. Monitoring of air pollutants was also undertaken at the centre of the room B (living room) at a height of 150 cm.

Carbon monoxide concentration was measured with a battery operated CO Personal monitor (model ecolyser-2000) regularly at an interval of ten minutes during one hour stove operation period. The CO concentration for all appliances was further measured till the CO value reduced to the prescribed values of air standards. The suspended particulate matter was collected on the glass fibre filter paper (GF/A) of 25 mm diameter by drawing air at a rate of 1.5 litres per minute. Blank filter and samples were kept in a desiccator and the mass of particulate matter was determined gravimetrically[12]. The suspended particulate matter from the filter was extracted with 150 ml benzene in a light protected solvent extractor for a period of 6 hours. The extract was concentrated to about 1 ml on a rotatory vacuum evaporator. The PAH were isolated by thin layer chromatography on a 2 mm preparative silica gel coated glass plates[13]. The PAH fraction was eluted in benzene for determining benzo (a) pyrene concentration by gas chromatography in a BBBT column[14].

TABLE 26.2: **Characteristics of the biofuels**

	Fuelwood	Dung-cake	Charcoal	Biogas
Fixed carbon	18.0	14.5	64.0	CH_4 62
Volatile content	70.0	60.8	23.0	CO_2 38
Ash content	2.0	15.3	2.4	
Moisture content	10.0	9.4	8.2	
Calorific value, kJ/kg	16,100	10,920	29,200	40,800*

* kJ/m^3

Results and Discussion

The concentration of carbon monoxide in indoor environment at the three positions during one hour combustion of acacia wood, dung cake, charcoal and biogas is presented in Table 26.3. For all the four biofuels, the indoor concentration of carbon monoxide increases linearly with time during the test. However, the rate of addition of CO in the indoor environment is quite different for different fuels. While the CO concentration in the beginning of the experiment are very close to each other towards the end of the experiment these values differ considerably. For example the CO concentration for the case of fuelwood is on an averge 8-17% higher than that for its charcoal. Similarly the direct combustion of dung cakes releases 98-134% more CO as compared to the combustion of biogas. Among the four biofuels, the dung cake combustion results in the maximum CO emission in the indoor environment whereas the processed biofuel obtained by the anaerobic digestion of the wet dung (i.e. the biogas) releases minimum CO on combustion in the biogas cookstove. The concentration of CO for the cases of acacia wood and its charcoal lie between the values for dung cakes and biogas combustion. The maximum and minimum concentration of carbon monoxide at two other positions i.e., standing position in room *A* and room *B* is also caused by the combustion of dung-cakes and biogas respectively. The concentration of indoor carbon monoxide was found maximum in room *A* at standing posture for all four biofuels. On other hand minimum concentration of carbon monoxide was observed at breathing level height in room *A* itself. The concentration of carbon monoxide in room *B* was found in between the two extremes.

The disappearance of carbon monoxide from the indoor environment after extinguishing the stove fire is presented in Table 26.4. It may be noted that minimum time required for reducing the level of CO from indoor environment is for biogas. The indoor CO concentration reduces to the prescribed WHO air quality limits after 20-40 minutes for different biofuels.

The concentration of suspended particulate matter in the indoor environment at the three positions is presented in Table 26.5. Once again as in the case of carbon monoxide, the maximum and minimum particulate matter was observed in room *A* at breathing level in standing posture and breathing level in squatting posture respectively. Combustion of dung cakes releases maximum suspended particulate matter (SPM) at all the three monitoring positions. The combustion of biogas leads to minimum SPM concentration among the four biofuels. It may be seen that a shift

TABLE 26.3: **Carbon monoxide concentration in the indoor environment during the stove operation period**

Position	Fuel	Concentration, mg/m³					
	Time, minutes	10	20	30	40	50	60
Room *A* Breathing level in squatting posture	Fuelwood	1	3	6	9	13	15
	Dung-cakes	2	5	9	12	14	18
	Charcoal	1	4	7	11	13	16
	Biogas	1	3	5	8	10	11
Room *A* Breathing level in standing Posture	Fuelwood	5	15	24	32	37	41
	Dung-cakes	7	18	27	35	44	52
	Charcoal	5	13	19	25	29	35
	Biogas	3	9	14	18	23	26
Room *B* Breathing level in standing posture	Fuelwood	4	10	15	17	22	27
	Dung-cakes	6	14	20	25	31	36
	Charcoal	3	8	11	16	20	24
	Biogas	2	5	9	11	15	17

from acacia wood to its charcoal reduces the particulate level by 50-65% at the three positions. Similarly a shift from dung cakes to the biogas reduces the particulate level by about 90% at the three positions.

TABLE 26.4: **Disappearance of indoor carbon monoxide after extinguishing the stove**

Time, minutes	Stove			
	Sugam-II (Fuelwood)	Sugam-II (Dung-cake)	Charcoal	Biogas
0	41	52	35	26
10	30	39	25	15
20	21	28	17	9
30	12	19	9	4
40	7	10	5	1
50	2	4	2	-

TABLE 26.5: **Concentration of suspended particulate matter in the indoor environment from four biofuels**

Fuel	Stove	Suspended particulate matter μg/m³		
		Room *A* breathing level in squatting posture	Room *A* Breathing level in standing posture	Room *B* Breathing level in standing posture
Fuelwood	Sugam-II	950	2450	1890
Dung cake	Sugam-II	1540	2600	2290
Charcoal	Charcoal	410	790	570
Biogas	Biogas Stove	105	250	195

A comparison of the concentration of suspended particulate matter observed in the present study with the air quality standard values shows that the combustion of Acacia wood, dung cake and charcoal releases particulate matter in the indoor environment higher than the prescribed limits. The particulate matter released from biogas is within the air quality standard values[15] (200-300 μg/m³).

The concentration of SPM released in the indoor environment by the combustion of fuelwood and dung-cakes observed in the presented study are similar to those reported by Raiyani *et al.* (1993)[16].

The concentration of benzo (a) pyrene extracted from the suspended particulate matter collected during the combustion of four biofuels is shown in Table 26.6. The benzo (a) pyrene emission like the two other pollutants is observed maximum in room *A* at breathing level in standing posture. While the minimum concentration was observed in the same room at squatting posture. In this case as well combustion of dung-cakes causes highest benzo (a) Pyrene emission at all the three monitoring positions.

Concluding Remarks

The present study shows that the combustion of processed biofuels releases lower concentration of pollutants as compared to their direct combustion. Thus the lower quality biofuels particularly the dung-

TABLE 26.6: **Concentration of benzo (a) Pyrene the indoor environment from the four biofuels**

Fuel	Stove	Benzo (a) pyrene, μg/m³		
		Room *A* breathing level in squatting posture	Room *A* Breathing level in standing posture	Room *B* Breathing level in standing posture
Fuelwood		0.280	0.610	0.480
Dung cake		0.415	0.970	0.670
Charcoal		0.135	0.240	0.205
Biogas		0.020	0.044	0.028

cakes should be converted to biogas which could be used more efficiently. This also provides nutrient rich slurry which may be used for increasing the biomass production. The inhabitants standing in the room *A* (i.e., room in which stove is operated) are exposed to the highest concentration of the pollutants as compared to the squatting posture. Thus the use of processed fuel should be encouraged.

NOTES

1. CMIE, Current Energy Scene in India, Centre for Monitoring of Indian Economy, Economic Intelligence Service, (1992).
2. Gerald, Leach and Gomen Marcia, Household Energy Handbook : An Interim Guide and Reference Manual, World Bank Technical Report No. 67, The World Bank, Washington DC, USA (1987).
3. World Health Organisation, Biomass Fuel Combustion and Health, WHO Int. DOC. EFP/84-64, World Health Organisation, Geneva (1984).
4. K. Sexton, R.D. Trietmann, J.D. Spengler and A.T. William, Characterisation of Indoor Air Quality in Wood Burning Residences, Environ. Int. 12, 265-278, (1986).
5. V. Joshi, C. Venkataraman and D.R. Ahuja, Emission Characteristics of Different Biomass Cookstoves, Pacific and *Asian J. of Energy*, 2 (4) 1 (1991).
6. J.B. Kandpal and R.C. Maheshwari, Estimation of Pollutants Released During Biomass Combustion, *Ind. J. Environ. Prot.* 11 (9), 693 (1991).
7. J.B. Kandpal and R.C. Maheshwari, Domestic Air Pollution Caused by Biomass Fuels, Presented in First Bioenergy Covention and Symposium (1991).
8. H.W. de Koning, K.R. Smith and J.M. Last, Bulletin of the World Health Organisation, 63, 11, (1985).

9. T.S. Patel, C.V. Raiyani, Mohan Rao, A.L. Aggarwal, G.P. Kartha, B.B. Chatterjee, Indoor Air Pollution Problems : Traditional Versus Modern Fuels, In Proc. of 3rd Int. Conf. on Indoor Air Quality and Climate, Vol. 4, 295-301, Swedish Council of Building Research, Stockholm, (1984).
10. J.B. Kandpal and R.C. Maheshwari, A Decentralised Approach of Biocoal Production in Mud Beehive Kiln, Bioresource Technol., 43 (2) 99 (1993).
11. P.S. Rajpal and R.C. Maheshwari, Problems and Prospects of Improved Biomass Stoves, Changing Villages, 10 (2) 1 (1991).
12. Indian Standard Method for Measurement of Air Pollution, Suspended Particulate Matter, Part IV (1974).
13. A.L. Aggarwal, C.V. Raiyani, P.D. Patel and S.K. Chaterjee, Assesment of Exposure to Benzo (a) Pyrene in Air for Various Population Groups in Ahemadabad, *Atmos. Environ.* 16 (4), 870 (1982).
14. J.M. Janini, G.M. Muschik, W.L. Zeilinski, N-N\- Bis P-Butoxy Benzyldiene a-a\ bi-p toluidene Thermally Stable Liquid Crystal for Unique Gas Chromatography of Polycyclic Aromatic Hydrocarbons, *Anal. Chem.* 48 (6), 809. (1976).
15. World Health Organisation, Air Quality Standards, Pollution in the Megacities of the World, (1992).
16. C.V. Raiyani, S.H. Shah, N.M. Desai, J.S. Patel, D.J. Parikh and S.K. Kashyap, Characterisation and Problems of Indoor Pollution Due to Cooking Stove Smoke, *Atmos. Environ.* 27 (1) 1643 (1993).

27

BEHAVIOURAL PROCESSES IN THE ADOPTION OF A NEW TECHNOLOGY

Case Study—Transfer of Improved Cookstoves in Village Wazirpur

—DILBAGH KAUR

Since the beginning technology has played a vital role in human development removing drudgery and generating additional income. Although a technology is being introduced for the betterment of human being and society, yet it is observed that it is not accepted immediately and whole-heartedly by the people. There could be innumerable factors affecting the human psychology for accepting a new technology. An humble effort has made to identify the psycho, and socio-cultural factors which enhance or inhibit the rural people to accept a new technology.

Case Study Area: Village Wazirpur District Gurgaon

The fact that owning products of new technology entails social status has a strong bearing on the acceptance of technology. The other factors being socio-cultural factors like the type of food they take, importance of season on the use of improved cookstove and availability of fuel. Old habits and the cultural ethnocentrism inhibit the women in their acceptance of cookstoves.

The village has 219 families/households. Mainly two castes are prevalent in this village: (1) Rajputs, (2) Scheduled Caste. Rajputs are land owners and fall into the higher and middle socio-economic group. Scheduled caste either work on the fields of Rajputs or do some other labour job like road construction. Their women also help them in the job. However, women folk of Rajput do not work outside their home premises. They may be educated yet not economically supporting their families. They usually pass their free time by sitting together in group and gossipping. On interviewing them, it was found that they are satisfied with this way of living.

Parameters of Improved Cookstove

The following parameters have been kept in mind to design the improved cookstove:

- — A design which gives maximum efficiency in terms of fuel consumption.
- — The byproducts should not be harmful and should not pollute the environment.
- — The smoke could be let out of the chimney so that it does not pollute the kitchen.
- — Saving in time while cooking and cleaning utensils.

Action Plan

A total of 40 cookstoves were installed in the village.

People's Responses

Three different reactions of the people were observed. These were:

1. An instant 'Yes'
2. An instant 'No'
3. Hesitancy on the part of few, but who after some thought said 'yes'

Through general observation and dialogues with them, it was noted that few of those who said 'yes' to new technology accepted it for the reason because it gives boost to their status and, therefore, is prestigious to have something new and different than others. And also, they can show off to others that they are modern. Most of the families who exhibited this behaviour belonged to the upper caste, few others did not actually understand its merits but just imitated others. Some of the families did understand the advantage of new technology and thought it to be better than the traditional cookstoves. It was found that persons, who are better-off economically, who have some formal education, who belong to upper caste, and who are doing jobs related to service are more aware of the improved cookstoves. All these factors are positively correlated with the use of these techniques.

TABLE 27.1: **Coefficient of correlation (r) between Demographic Variables; and Awareness, Use, Reason and Adoption of Improved Cookstove**

S.No.	Variables	Awareness (r)	Use (r)	Reasons (r)	Adoption (r)
1.	Age	-0.104	-0.030	0.156*	0.015
2.	Caste	0.205**	0.201**	0.267**	0.352**
3.	Family Structure	0.116	0.161*	0.039	0.142*
4.	Education	0.162*	0.189**	0.077	0.202**
5.	Occupation	0.223**	0.148*	0.258**	0.344**
6.	Economic Status	0.268**	0.358**	-0.013	0.263**

** Significant at 0.1 level
* Significant at 0.05 level

While stating the reasons for using the improved cookstoves, caste and occupation plays the predominant role, indicating that persons belonging to upper caste and doing service could state more valid reasons for using the improved cookstove. Findings also indicate that, higher the age, greater is the knowledge of the improved cookstove. The ultimate adoption (awareness, use and reason) of the improved cookstove has significant correlation with caste, education, occupation and economic status, all of which are significant at 0.001 level of confidence. Persons belonging to joint family also adopt the improve cookstove, because of its time efficiency, as more work has to be done in less time.

For feedback, initially monthly and then seasonal visits were arranged to the village. It was seen that after a year's time, most of village people had reverted back to the traditional cookstoves. Thus three categories of people could be identified—(1) who have not accepted the improved cookstoves from the very beginning, (2) who have accepted it for a short duration (less than a year) and (3) who have accepted it for a slightly longer duration (up to one year).

Possible Reasons for Non Acceptance of Improved Cookstoves

(i) Dialogues with the people revealed that the non-acceptance shown by the people was because of their suspicion of the motives and intentions of persons from the city and government agencies. This feeling may be stemming from their experience of centuries of colonialism.

(ii) Too much intervention of government and other voluntary agencies for rural upliftment have made rural people immune to the development programme.

(iii) Besides this, attitudes of rural people towards government schemes did not help in expansion. Their attitude is to go for this change only with government aid and they did not want to take any enterprise of their own. Most of them who did not accept the new technology belonged to lower socio-economic group.

(iv) An attitude of conservatism and cultural ethnocentrism also inhibited the villagers from using the new technology. For them whatever they have is the best. This makes them hold to their traditional cookstove, and prevents them to go for a change.

(v) Relative value of the planned innovations have an important role in accepting any technology. While taking the regular feedbacks from the people who had installed improved cookstoves, it was realised that many of them stopped working on improved cook stoves and come back to their traditional stove. For them, the relative value of other reasons is more strong than the reasons for which they have installed cookstoves. Due to the cultural pattern, second pot of the cookstove is not used during summer season, so there is unnecessary wastage of fuel. If the chimney breaks then it loses its concept of smoke reduction, that means extra expense to put another chimney. These kinds of relative values defeat the very purpose for which it was introduced.

(vi) Unforeseen consequences of the planned innovation. There are always intended consequences of a certain research/ technology. Here we know that the use of improved cookstove increases efficiency, removes drudgery of rural women and saves fuel and time in cooking. At the time of innovation, it was not thought that the smoke from the traditional cookstove removes the termite of the wooden roof. If the smoke is eliminated then white ants infest the roof, that means, the replacement of another roof which is much more expensive than the benefits of the improved cookstove.

(vii) Besides villagers are so habitual to their traditional cookstoves that they did not want to work on the new cookstove.

(viii) Also it is a cultural pattern in the rural houses to break and build the cookstove off and on. This they could not do with the improved cookstoves as they were not at ease to rebuild it, in spite of the fact that the new cookstove was similar to the old one and easy to make.

(ix) Finally while rebuilding, the dimensions of the cookstoves were not kept in mind which reduced the efficiency of the cookstove.

Reasons for Rejection

Regular visits revealed that many of rural people who initially had accepted the improved cookstoves are not using it any more. Reasons of rejection could be many.

(i) Two major improvements over the traditional cookstove, i.e., fuel efficiency and time saving did not mean much to them. They were not facing fuel crisis as they used to burn dry twigs which are abundantly and freely available throughout the year in that area. Thus it hardly mattered to them whether the fuel consumption was more or less.

(ii) Also it was seen that out of 40 only 4 houses appreciated the time element. It was because the housewives had plenty of leisure time and they had nothing else to do other than household chores. It was only in those four houses where men had to go to far off places for their job; that early morning meals were of great importance to the family and thus they realized and appreciated the time efficiency of improved cookstove.

(iii) Food habits also played an important role in the acceptance of the cookstove. The housewives who opted for the new cookstove because of its two pots realised that they mostly ate a cereal and a pulse/vegetable, thus making the second pot useless. A survey showed that only five houses make cereal, pulse and vegetable. Ten families ate cereal and pulse or vegetable. While the rest only ate cereal (with milk and salt).

(iv) During winters, all the forty houses were using both the firebox irrespective of their food habits. Most of the time, the second firebox was used to boil the milk, make tea or heat water for bathing, washing clothes or utensils. But in summers, not much use was found for the second firebox.

(v) Cookstove are mainly built in open courtyard but it is problem to work on it in the rainy season, summer afternoons and winter evenings. The new technology, being a permanently fixed and built cookstove, could not be shifted conveniently.

(vi) During the visits, it was also observed that many of the rural people were attracted by the author as a person, due to her casual and informal behaviour. This was relayed to the author in words also. Author thinks that families who instantly said 'yes' to the new technology, also could be due to identifying themselves with the person. They perhaps accepted the cookstove to please her rather than to use the new technology with its improved benefits over the traditional one.

(vii) Through feedback, it was noticed that many of the families did not understand the utilisation of the parts like dampers and chimney, and also the measurement and the distance between the two fire-boxes. Often they remove the dampers or the chimney. This kind of behaviour defeats the objective of efficiency in saving fuel and smoke reduction. Also when they reconstruct it, they do not care for the measurements, which then does not gives the desired results.

CONCLUSIONS AND RECOMMENDATIONS

National Bio-Energy Convention-95

Mission Approach for Bio-Energy Development Necessary

The bio energy resources need to be developed in a mission mode involving various ministries, departments and agencies as these resources are crucial in meeting the rural energy needs by making energy available in an equitable fashion and thereby becoming a major tool in rural development and employment overcoming rural poverty.

A policy Framework and an Active Plan for Energy in rural areas is lacking. A national plan should be immediately developed. The responsibility for coordination of all activities relating to bio-fuel production, conversion and utilization should lie with one agency. The emphasis of such a plan should be on finding an urgent solution to subsistence level agriculture and human drudgery and on consolidation of efforts for achieving rural development through income generation activities, leading to better food security, health and sanitation.

Emerging village Panchayat mechanism was strongly recommended for planning and implementing bio energy programmes in the rural areas. People's participation and women's participation were strongly recommended.

Recent technological advancements in high temperature gasification; and the need to dispose of increasing urban, municipal and industrial waste makes bio energy options necessary also for the urban centres providing energy and organic manure at the same time.

Recent observations of micro level fuelwood production have shown stabilisation of fuelwood prices. The energy plantation should be vigorously pursued so that our bio-diversity is appropriately harvested. Various issues like land-use, bio-diversity patenting, optional uses of fuelwood species and financing and managing of projects should be given due consideration.

These views were expressed by the participants to the Ninth National Bio-Energy Convention, with the theme Bio-Energy for Rural Energisation concluded here, and organised by the Bio-Energy Society of India and IIT, Delhi. The Convention was attended by a large number of bio energy experts, officials, foresters and wasteland development experts.

The participants made following specific recommendations:

1. The Bio-Energy resources have a crucial and relevant role in supplementing the meagre per capita availability of energy in rural areas (as these are available in a decentralised and equitably distributed fashion). Higher utilisation of bio resources should be a national priority.
2. A National Plan for Rural Energy should be developed on priority, with due emphasis on bio energy resources, as these only provide low cost decentralised option for the poor populations. The reflection of bio energy resources should be reflected in the Ninth plan.
3. A Mission Mode Programme for bio energy development and use aimed at better rural development through income generation opportunities should be launched without losing any time. The Mission should involve various agencies and have a time bound programme for achievement by 2000 A.D.
4. The emerging third level governance of village panchayats should be appropriately and effectively used as a delivery mechanism for providing the support systems for the youth in the rural areas to adopt bio-energy based systems for income generation.
5. Urban municipal and industrial waste disposal has become a major problem in all cities. Immediate efforts be launched for setting up systems that will convert these wastes into energy and organic manure. Sustainable operation and maintenance of such systems shall call for private and NGO involvement.
6. Wastelands Development Department, besides being involved in growing energy plantations on wastelands, should also realise the complementary role of efficient energy conversion devices and should, therefore, undertake

programmes of implementation of high efficiency chulha/ gasifier, briquetting machines, turners for rural industries.

7. Micro level resource assessment of agro residues should be taken up on priority, as major efforts in India on gasification and briquetting routes are facing major set-back due to lack of sustained raw material availability so important for emerging small scale industries.
8. Need based, site specific high productivity species should be identified, developed and disseminated on a large scale. In order to provide quality planting/seed material, nurseries must be supported. Training on nursery techniques should also be organised.
9. Bio-fertilizers are decentralised source of plant nutrient for biomass production complementing chemical fertilizers, thereby conserving soil productivity and health and saving in fossil energy (needed for production of chemical fertilizers). Effective use of bio-fertilizers should be achieved by promoting integrated plant nutrient supply and management system.
10. Appropriate fund allocations should be made for reconditioning and replacement of bio-energy conversion systems installed in past.
11. Land-use legislation conducive to use of forest and public lands for energy plantations should be promoted.
12. Awareness should be created about patenting of species of fuelwood trees and related technologies.
13. Awareness creation programmes aimed at highlighting the role of energy plantations for fuel, food, fodder and bio fertilizers and fibre should be promoted.
14. Awareness about bio-energy resources should also be promoted through the school education text books so as to create an impact at a young age.
15. People's participation through involvement of local NGOs and women's organisations should be promoted.
16. Emphasis should be given on R&D for conservation, conversion and utilization aspects upto end users.
17. Industries should be promoted to take lead in R&D and development projects and cost effectively producing devices and systems of bio-energy resources. The Government policy should support empowerment of

industry to take decisions and operate under the market forces only.

18. Training programmes aimed at entrepreneurship development in technical, financial and management practices should be promoted for energy plantations, conversion and utilization sectors.
19. In view of the recent successes in quantifiable production levels, alternative uses of traditional fuelwood species should be developed so as to make energy plantation viable under private entrepreneurship.

AIMS AND OBJECTIVES OF BIO-ENERGY SOCIETY OF INDIA

1. NAME OF THE SOCIETY

   The Society shall be called Bio-Energy of India. The name of the Society in Indian language shall be Bhartiya Jaiv Urja Sanstha.

2. REGISTERED OFFICE

   The Registered Office of the Society shall be situated in the Union Territory of Delhi. At present, it is situated at Indian Institute of Technology, Hauz Khas, New Delhi-110 016.

3. AIMS AND OBJECTIVES

   (i) to promote the development, spread and utilisation of bio-energy with particular emphasis on meeting rural energy needs.

   (ii) to increase public understanding on the potential, production, conversion, utilisation and conservation of bio-energy maintaining ecological balance, and environmental protection by various means including dissemination of information through the mass media.

   (iii) to promote and undertake study and research in various aspects relating to bio-energy.

   (iv) to assess scientific and technological advancements in various countries, the problems and possible solutions in the Indian context.

   (v) to uplift the living standards of the people through utilisation of bio-energy resources.

   (vi) to organise conferences, seminars, lectures, exhibitions and study groups of local, regional, national and international levels.

   (vii) to establish and maintain relations with similar bodies in other countries.

LIST OF CONTRIBUTORS

K. Balakrishna Reddy
Indian Renewable Energy Development Agency, Core-4A, East Court, India Habitat Centre, Lodhi Road, New Delhi.

R. Balasubramaniyan
Central Power Research Institute, Energy Research Centre, Sreekariyam, Thiruvananthapuram, Kerala.

A. Chakraverty
Agricultural and Food Engineering Department, Post Harvest Technology Centre, IIT, Kharagpur, India.

Pradeep Chaturvedi
Secretary General, Indian Association for the Advancement of Science, New Delhi.

Suneeta Chopra
Energy Research Centre, Panjab University, Chandigarh.

P.D. Grover
Biomass Conversion Laboratory, Department of Chemical Engineering, IIT, New Delhi.

T.M. Haridasan
Action Research Centre on Biomass Gasification School of Energy Sciences, Madurai Kamaraj University, Madurai.

P.V.R. Iyer
Biomass Conversion Laboratory, Department of Chemical Engineering, IIT, New Delhi.

Dilbagh Kaur
Centre for Rural Development and Technology, IIT, New Delhi.

J.B. Kandpal
Centre for Rural Development and Technology, IIT, New Delhi.

Vikas Kumar
Sardar Patel Renewable Energy Research Institute, Vallabh Vidyanagar, Gujarat.

S.B. Kute
Additional General Manager (AD&AS), Gujarat State Fertilizers Company Ltd., P.O. Fertilizer Nagar, Distt. Vadodara, Gujarat.

Banwari Lal
Agro-Energy Unit, Division of Agronomy, IARI, New Delhi.

Mira Madan
Centre for Rural Development and Technology, IIT, New Delhi.

J.N. Maggo
Joint Adviser, Planning Commission, New Delhi.

R.C. Maheshwari
Centre for Rural Development and Technology, IIT, New Delhi.

Nisha Menon
Indian Renewable Energy Development Agency, Core-4-A, East Court, India Habitat Centre, Lodhi Road, New Delhi.

S.K. Mishra
Biomass Conversion Laboratory, Department of Chemical Engineering, IIT, New Delhi.

A.K. Mukerjee
Former Inspector-General of Forests and Vice-Chairman, National Afforestation and Eco-Development Board, M/o Environment and Forests, Government of India.

M. Murugan
Action Research Centre on Biomass Gasification School of Energy Sciences, Madurai Kamaraj University, Madurai.

G.P. Nagori
Sardar Patel Renewable Energy Research Institute, Vallabh Vidyanagar, Gujarat.

B.S.K. Naidu
Indian Renewable Energy Development Agency, Core-4-A, East Court, India Habitat Centre, Lodhi Road, New Delhi.

V. Narasimhan
Action Research Centre on Biomass Gasification School of Energy Sciences, Madurai Kamaraj University, Madurai.

I. Natarajan
Economic Adviser, National Council of Applied Economic Research, New Delhi.

B.J. Patel
Sr. Manager (Bio-fertilizer), Gujarat State Fertilizers Company Ltd., P.O. Fertilizer Nagar, Distt. Vadodara, Gujarat.

S.R. Patel
Sardar Patel Renewable Energy Research Institute, Vallabh Vidyanagar, Gujarat.

R. Perabagaran
Managing Director, Chennai Dendro-Tech, Private Limited, Madras.

Rajendra Prasad
Centre for Rural Development and Technology, IIT, New Delhi.

B. Raja
Action Research Centre on Biomass Gasification School of Energy Sciences, Madurai Kamaraj University, Madurai.

P.S. Rajpal
Department of Applied Mechanics, IIT, New Delhi.

N.S. Rathore
College of Technology & Agricultural Engineering, Udaipur, Rajasthan.

N. Saha
Sardar Patel Renewable Energy Research Institute, Vallabh Vidyanagar, Gujarat.

S.K. Sharma
Energy Research Centre, Panjab University, Chandigarh.

Satyawati Sharma
Centre for Rural Development and Technology, IIT, New Delhi.

Pawan Sikka
Department of Science and Technology, IIT, New Delhi.

C.B. Singh
Agro-Energy Unit, Division of Agronomy, IARI, New Delhi.

Jagpal Singh
Centre for Rural Development and Technology, IIT, New Delhi.

H.K. Singh
Agro-Energy Unit, Division of Agronomy, IARI, New Delhi.

Narayan Singh
Former Additional Inspector General (Forests). Government of India.

Pratap Singh
College of Technology & Agricultural Engineering, Udaipur, Rajasthan.

D. Subhakar
Energy Research Centre, Central Power Research Institute, Sreekariyam, Thiruvananthapuram, Kerala.

G.S. Sekhon
Department of Applied Mechanics, IIT, New Delhi.

N.S.L. Srivastava
Central Institute of Agricultural Engineering, Bhopal, 462038.

G.C. Shrotriya
Senior Manager (Bio-Sciences), IFFCO, New Delhi - 19.

K.S. Shishodia
Department of Applied Mechanics, IIT, New Delhi.

M. Sudarsan
Central Power Research Institute, Energy Research Centre, Sreekariyam, Thiruvananthapuram, Kerala.

G. Thirumurugan
Action Research Centre on Biomass Gasification School of Energy Sciences, Madurai Kamaraj University, Madurai.

Joseph Thomas
Deputy Director, Shri A.M.M. Murugappa Chettiar Research Centre, Madras.

Padman Vasudevan
Centre for Rural Development and Technology, IIT, New Delhi.

V.K. Vijay
Centre for Rural Development and Technology, IIT, New Delhi.

R.K. Wanchoo
Energy Research Centre, Panjab University, Chandigarh.